Newlywed COUPLE'S DEVOTIONAL

52 Weeks of Everyday Scripture, Reflections, and Prayers for a God-Centered Marriage

Chris and Jamie Bailey

Zeitgeist · New York

Published in the United States by Zeitgeist, an imprint of Zeitgeist™, a division of Penguin Random House LLC, New York.

penguinrandomhouse.com

Zeitgeist™ is a trademark of Penguin Random House LLC

ISBN: 9780593196670
Ebook ISBN: 9780593196953

Cover and interior art © kate_sun/Getty Images
Book design by Katy Brown

Printed in the United States of America

1 3 5 7 9 10 8 6 4 2

First Edition

This book is dedicated to the biggest inspirations behind us in our own God-centered marriage—our girls, Alicia, Taylor, and Mackenzie.

Table of Contents

Introduction

CONGRATULATIONS—YOU'RE MARRIED! Whether you're just getting back from your honeymoon or hoping to relive it, we believe you're already on the greatest adventure of your life.

When we first said our vows over 25 years ago, we were far from God. We had no clue what it meant to follow Jesus or to make him the center of our marriage. This made for some rocky first years. We had our fun, but we had no clue how to love each other sacrificially or thrive in a godly marriage. By default, we both looked to marriage as a way to receive—to be completed by each other.

When we came to know and follow Jesus, everything changed. It became painfully clear how the vows we took together stood on sand, vulnerable to washing away. When we decided to make God's way our way, we began to follow the path he laid out before us to create the firm foundation we needed. Within these pages, you'll find biblical truth to lay this new foundation, too. These truths outlast time and require sacrifice, honesty, and obedience. They call you to love each other as Christ loves his church. They serve as the anchor for your marriage, as teachings that will both challenge and encourage you, and as practical tools to keep Christ at the center.

Your marriage is a journey filled with both hills and valleys. As you adventure together, we hope that God's Word and these practical devotions—written by us both as individuals and collaboratively, and containing many examples from couples we've counseled—will be the navigation tools you need to stay the course!

May it be his grace and love that guide you on your journey to building a God-centered marriage.

How to Use This Book

WE HAVE PROVIDED 52 devotions designed to be read together weekly over the course of one year. Every topic is rooted in Scripture and highly relevant for newlyweds and all married couples. Each week begins by digging into Scripture for a time of Bible study. Then, there are five discussion questions to help you go even deeper into each lesson. At the end, you'll find a specific, everyday prayer relevant to that week's topic, allowing you to ask God to help you apply his truth to your marriage.

As you get started on your journey, here are some practical tips to help you make the most of your time together:

- It may help you stay committed if you choose a day and time in advance to make this a part of your weekly rhythm. Find a routine that works for you.
- Be careful not to rush through each devotion. Take your time in reading the Scripture together and discussing it. Allow it to really settle into your hearts, keeping in mind that Scripture contains power for your marriage: "For the word of God is living and active, sharper than any two-edged sword" (Hebrews 4:12, ESV). Give it time to work. Then, plan a later time during the week to use the discussion questions to check-in with each other.
- Take it a step further. Grab a journal to write out any thoughts that come to mind as you ponder the verse for the week. Perhaps write out the Scripture and memorize it together. Find meaningful ways to apply it to your lives as you "let the word of Christ richly dwell within you" (Colossians 3:16).

- Be prepared for some resistance. There may be triggers that come up with sensitive topics, and you will also likely face opposition from the Enemy. Keep in mind that God has enabled each of you to "resist the devil" as you strengthen your marriage (James 4:7b).
- Lastly, be willing to rely on grace and flexibility for those hard and busy days. You want this time to be a positive, prioritized experience. Use your time together to grow, connect, and be vulnerable, allowing God to have his place in your marriage—right in the center.

Now you're ready! It's time to get started on your journey to becoming the best husband and wife you can be. Every week of commitment you put in will surely result in the God-centered marriage you're after.

A God-Centered Marriage

MOST FOLLOWERS OF CHRIST have an understanding of what it means to live a God-centered life—to seek God with their whole hearts and live according to what pleases him (Jeremiah 29:13). Now, as a newlywed couple, you must learn what it means to become a God-centered spouse. But don't worry, because we're here to help show you the way!

What God Says

Marriage is a union of two individuals, merging as one: "And they shall become one flesh" (Genesis 2:24). Bound by a covenant under God's design, we are instructed to "let not man separate" what God has "joined together" (Matthew 19:6). The marriage you are in right now is the marriage God intends for you to stay in. Your marriage has a purpose.

Throughout God's Word, the husband is a symbol of God, and the wife is a model of his Church. How a husband treats his wife is designed to reflect how Jesus treats his bride—the Church. Your marriage is meant to put God's love on display. Earlier on in our private practice, a younger couple we worked with, Bethany and Bryce, showed this clearly in their marriage. Bryce is lovingly protective, constantly looking out for Bethany's best interest. Whenever stress with her job arises in her life, he is right there, encouraging her to rest after work, while he steps in and helps. He deliberately gives her time to take a bath and relax while he cleans up after dinner. He is intentional about helping her safeguard her time and energy. As a result, Bethany values his opinions and accepts his guidance; you can see the true rest and peace she finds in trusting that Bryce will always be there for her.

The mutual love and respect they have for each other is obvious to all who know them.

When you seek to have a God-centered marriage, you are seeking to represent his character together. A God-centered marriage will be filled with selflessness, sacrifice, forgiveness, patience, joy, and perseverance. God never gives up on us, and so we should never give up on each other.

Being a God-Centered Spouse

Becoming a God-centered spouse is one of the highest callings in marriage, but what does that even mean? The biggest difference between being God-centered and spouse-centered is who you seek to please. A God-centered spouse will always seek to please God, not people. The Bible makes it clear that seeking to primarily satisfy others prevents us from serving Christ, which is the ultimate goal of a God-centered spouse (Galatians 1:10).

To be spouse-centered is to be focused on yourself and your spouse. Without God in the center, this focus on your spouse is not a devoted, sacrificial attention; it's a mentality that says, "I'll scratch your back if you scratch mine" and "I'll do my fifty percent if you do your fifty percent." It's a conditional love that places requirements upon your partner, and that is not the love God calls us to. He calls us to give 100 percent, whether or not your spouse gives in return. He calls us to a God-centered love that leaves room for grace, humility, forgiveness, and conviction from the Holy Spirit. Spouse-centered love, on the other hand, can leave room for pride and blame, which can lead us down the road to division.

As counselors, we have seen many struggling couples fall prey to a spouse-centered marriage. We remember when Mark and Ashley first came into our office. They were constantly fighting and caught in the blame game. Neither of them sought to do what was best for their marriage or what God deemed as right. Instead, they wanted to get their own way by changing their spouse. Mark refused to take any ownership for his volatile behavior and angry outbursts, because he felt justified by how Ashley spoke to him. Ashley wasn't about to change either, because she didn't deserve to be treated the way

Mark treated her. They were stuck. They were trying to please themselves over pleasing God. Fortunately, we knew exactly how to help them make that shift from seeking to please themselves to learning to please God. As a result, they're happier than they've ever been.

As newlyweds, bringing this God-centered perspective into your marriage serves as an invaluable foundation. Why? Because we get good at what we practice. If you enter into your marriage seeking to honor God first, then you'll practice becoming more and more like Jesus, and you'll get good at it. God loves when we are "doers of the word," and not hearers who deceive themselves (James 1:22). His Word and his ways are meant to be lived out and actively followed. God-centered marriages fulfill his design for what marriage means.

Being a God-Centered Couple

When you said, "I do," each of you committed to becoming a God-centered spouse. Those vows represented your commitment to each other and to God that your marriage would honor him. Unfortunately, maintaining this promise won't always be easy. Life will throw distractions your way, struggles will arise, and, if we're honest, attitudes can change on a whim. The way you felt when you first made this commitment may be a long way off from where you stand now.

We want to provide you with a framework for the best start you can have for your God-centered marriage journey. To begin, here are some tips you can put into practice right away, as you read and apply all that God has for you as a couple which will help as you read these pages:

- First things first, make an agreement with each other. Decide right now to take this journey together. Marriage is an all-in deal. Take a moment and commit to completing this 52-week journey and staying the course.
- Be ready for better and for worse days. Accomplishing the week's lesson will be harder on some days than on others. On these days,

do one of two things: be willing to walk in grace, or be willing to fight through, using the devotional to help transform the worse into the better.

- Focus on the goal of making this a holy matrimony. Go back to the wisdom of James 1:22 and be a doer of what you learn. Don't just read it and deceive yourself into thinking that's enough. Be intentional in asking God how you can make his Word come alive in your marriage.
- Start your day remembering that you vowed to forsake all others and be faithful to each other. Set your intentions each morning. Ask God to go before you, help you prioritize each other, and transform you into the best spouse you can be today.
- Lastly, seek to cherish each other. Remember the gift of the cross that Jesus gave for each of you. That love was poured into you and will overflow to your spouse if you allow it to. Each day, say, "I love you," and choose one act, even something as simple as making the coffee, to let your spouse know you mean it.

"As much as it is up to me, I vow to help make you more like Christ every day"—those are words from our oldest daughter's wedding. And though they won't always be easy to fulfill, we hope you give them your greatest attempt, too.

Worth Fighting For

"I have fought the good fight, I have finished the race,
I have kept the faith" (2 TIMOTHY 4:7).

NOW THAT YOU'RE MARRIED, it's essential that you learn how to be a good fighter. No, not so you can win battles against your spouse, but so you can fight for them and for your marriage. As the apostle Paul neared the end of his earthly life, there were three things he recognized as important: fighting the good fight, finishing the race he began, and keeping the faith. This involved persevering through all of his trials, continuing on with the assignment God gave him, and doing so with complete faith in God's plan. We want for you to embrace all three in your marriage, and, one day, to make the same claims Paul did at the end, that you kept going and finished strong.

Too often and too soon, couples give up on their marriages. That's just not a statistic you need to be a part of. All of us enter into marriage having no idea what to expect. We may think we know what it will be like, but let's be real—our preconceptions usually involve amazing communication, reciprocated kindness, and all the blissful sensations love stirs in us. There's nothing wrong with those ideals; in fact, they likely helped you decide to get married in the first place. And those things will indeed be present in your marriage, too, but there will be times you'll need to fight for them. We want you to be ready for that battle.

You have begun your journey of marriage, and it will be exciting and adventurous, but it will also be a place of growth and learning. Be eager and ready for those parts of the journey, too. Remember that your marriage has a purpose, as part of the race God has set before you, to help you become more and more like Jesus. It's worth fighting for, finishing, and fulfilling in truth. As you move forward, just one day at a time, enter each moment as a warrior willing to fight for the marriage that God blessed you with. Your main job as a spouse is much like Paul's as a follower of Christ—to fight the good fight, finish the race, and keep the faith. If you commit to doing these three things in your marriage, God will be sure to do his part as well.

1. Discuss your thoughts on how you each might handle hardship or unexpected trials in your marriage. Can you come up with a game plan now in order to be better prepared? Try coming up with example scenarios, perhaps by examining what your friends or parents have experienced in their own marriages.
2. Ask each other, "What does me being willing to fight for you look like?"
3. Is there an area now where you need my help fighting for you—a place where I can stand up for you or let you know I've got your back?
4. What do fighting the good fight, finishing the race, and keeping the faith in our marriage mean to you?
5. Take a minute to hold hands, look at each other, and say these words: "I have your back, and I promise to always fight for you and for us."

Dear Lord,

Thank you for giving us each other, and for creating a partner in life who we can fight for. Please give us the endurance we need in order to stay the course that you have planned out for us. Help us to always have the desire to run our course together, and the ability to do so faithfully in truth.

In Jesus' Name, Amen

Cutting Grass and Doing Laundry

"For as in one body we have many members, and the members do not all have the same function, so we, though many, are one body in Christ, and individually members one of another" (ROMANS 12:4-5).

OFTENTIMES, couples enter marriage with an automatic assumption of what doing the chores and household duties will look like. You might believe you're on the same page about who will mow the lawn or wash the clothes, but you'll find out quickly if you're not. Of course, we don't mean to over-spiritualize household chores, but honoring each other's design matters when making a functional, God-centered home, even when it comes to the small stuff.

In the book of Romans, the author Paul makes it clear that we're not carbon copies of one another. We don't possess all of the same gifts, strengths, or abilities, and that's a good thing. Embracing our differences helps the Church function at its best—as a body with every part represented. This truth applies for your marriage. The person gifted in service, who volunteers to stack chairs on Sundays, may not feel compelled or equipped to preach at the pulpit. Likewise, the spouse who's free-spirited and unstructured shouldn't have to plan the schedule or ensure the bills get paid on time if the other spouse excels at organization. No task or gift signifies greater importance or higher status than another; they're all equally vital in forming the body of Christ. Abandoning your assumptions and working as uniquely individual members of one unit will help your home function at its best.

In Brad and Melissa's marriage, for instance, Melissa is more outgoing and active. Around the house, she likes to mow the grass and wash the cars. On the other hand, Brad is more reserved and meticulous. He likes to do the laundry and stay on top of the bills. This division of chores works best for their home, but it may not be what works best for yours. Ultimately, it's up to you to decide what division of duties will help your home thrive. This is not about stereotypical roles; it's about communicating and embracing our personalities and the gifts we were given.

1. What assumptions did you bring into your marriage about who would do what chores in your home?
2. Who took care of which chores in the home you were raised in?
3. Is there a way that God has gifted each of you that makes more sense for specific tasks?
4. What is your current attitude about doing work around the home? Do either of you have concerns about this area based on your answers from questions 1 and 2?
5. Together, make a list of all the important tasks that need to be done in your home. Include even the minor or mundane ones, so that they don't get overlooked or become an issue down the road. Individually, identify the top three chores from this list that *you* would like to do and the top three chores you would like *your spouse* to do. Share them with each other. Go down the list you've created. Beside each chore, write the initials of the spouse who will complete it. It's okay if you put both of your initials beside a lot of them!

Dear Lord,

Thank you for giving us each our own set of unique strengths and gifts. Help us to use those strengths to run our house as you see fit. Give us hearts that appreciate each other, and let us enjoy the distinct way you have created each of us.

In Jesus' Name, Amen

You're the Best!

*"And a voice came from heaven, 'You are my beloved
Son; with you I am well pleased'"* (MARK 1:11).

WHEN JESUS WAS BAPTIZED, before any of his documented earthly ministry
or miracles had occurred, God saw fit to remind him how pleased he was
with his Son. In this moment, God wasn't proud of what Jesus had done;
he was proud of who he was. I can only imagine the face of Christ as those
words spoken from heaven reached his ears. I imagine the look on his face
resembling my husband's own expression when I publicly remind him how
appreciative I am of who he is.

If Jesus appreciated words of blessing, and if God found them important
enough to demonstrate for us, then we should take note. Every single one of
us longs for affirmation, to know that we've made someone we love proud
of who we are. Hearing words of blessing and validation impacts our hearts
and reminds us that we have value, that we are seen—especially when those
words come from someone we love.

Everyday life together can easily become mundane. We establish rou-
tines that take us from place to place, chore to chore, and job to job without
interruption. In general, life has a way of taking more from us than it gives
back, and it's easy to grow weary and worn-down. This makes it all the more
critical to slow down and speak words of life over each other. What a gift of
encouragement God gave his Son that day. God knew all that would soon
unfold for Jesus, and he chose to propel him forward, reminding him how
satisfied he was with him from the very start. We may not know what lies
ahead for our spouse, but, as husbands and wives, we have the ability to equip
one another with the kind and thoughtful messages we speak. There's not

one day we can't face a little more bravely and optimistically, knowing someone is already proud of us.

Make words of blessing a priority in your marriage. Set aside time to thank your spouse. Remind them of how great you think they are and how much you appreciate all that they do.

1. When was the last time you received praise or a blessing from someone? This could be a parent, a coworker, a friend, and so forth. What did they say?
2. When was the last time you spoke a blessing of praise or gratefulness over each other?
3. Did you grow up receiving more praise or more criticism? Discuss what this was like.
4. Is it easy for you to receive praise or compliments? Why or why not?
5. Get your phone out and set a timer for 60 seconds. Each take a minute to tell the other all of the traits you love about them.

Dear Lord,

Thank you for giving us each other. Help us in reminding us how much we mean to each other and in recognizing the other's worth. Open our eyes to the times when our spouse needs encouragement from us the most. Assist us in using our words that we have for each other as the exact encouragement we each need in our daily lives.

In Jesus' Name, Amen

Naked and Unashamed

*"And the man and his wife were both naked
and were not ashamed"* (Genesis 2:25).

WHEN GOD CREATED the first husband and wife, Adam and Eve, he designed complete transparency between them; they were naked and unashamed. This doesn't just signify their physical nakedness, but full spiritual, emotional, and mental exposure—an all-in, nothing-to-hide nakedness. This was a beautiful gift to be shared between a husband and wife, rendering each of them completely raw and vulnerable—fully seen, known, and loved by God and each other.

As we know, things have changed since then. With the Fall, shame entered into our lives and, subsequently, invaded our marriages. It persistently nags us with the message of "I am not enough." It intrudes, creeping in without warning and compelling us with a need to hide, even from each other. It demands that we find our fig leaves and cover up, body and soul.

When we first met Jennifer and Tom, Jennifer constantly struggled with not feeling good enough or pretty enough. This shame affected her so pervasively that she began to hide from Tom. She didn't tell him her fears and insecurities, and she was even uncomfortable allowing him to see her undressed. Meanwhile, Tom struggled all on his own. Jennifer's emotional withholding, along with her physical withdrawal, left Tom feeling like he was not enough. Maybe his job didn't earn enough to make her happy, or she wasn't satisfied with him in the bedroom. Perhaps the problem was him. As their counselors, we addressed each of these layered concerns, knowing when shame enters a marriage, it can paralyze both spouses and keep them from growing.

If you ever feel shame creeping in, it's time to go back to the beginning and seek restoration of God's original design for marriage. A restored marriage is rich in vulnerability and trust, with partners who are freely giving their love and acceptance to each other. Within it, husbands and wives embrace their flaws, learning to love each other deeply in spite of them. It's time to tell the accuser, deceiver, and shame-bringer that he no longer has room in your marriage. Drop the fig leaves, lean into exposure, and allow yourself to be fully seen and fully loved.

1. What do you think it would be like to have the kind of freedom that Adam and Eve originally had in the Garden of Eden?
2. Take turns sharing one of your insecurities with your spouse. When it's your turn to listen, respond only with, "Thank you for sharing that."
3. Do you tend to struggle more with physical or mental insecurities? In other words, do you struggle with body image and your looks, or do you struggle with your intelligence, ability to fit in with your peers, being good enough at your job, and so forth?
4. Describe an insecurity that you've had most of your life. When do you think it first started? How does it still affect you?
5. How comfortable are you standing physically naked before your spouse? If this is something you would like to change, start small by undressing to whatever degree you are comfortable with in dim lighting. To progress, either gradually increase the lighting or challenge yourself to remove more clothing each week.

Dear Lord,

We ask that you keep shame, which makes us doubt how you made us and how we truly feel about one another, from entering into our marriage. Help us to be courageous enough to be vulnerable with each other, both physically and emotionally.

In Jesus' Name, Amen

If You're Wrong, You're Wrong

*"So then let us pursue what makes for peace and
for mutual upbuilding"* (ROMANS 14:19).

WE'VE GOT SOME bad news for you. You're not going to be able to have a marriage without some mess-ups. And, where mess-ups exist, so does the need for apologies—and apologizing can be difficult. Yet, without them, we experience breaks in connection and a lack of peace. So how do we pursue what makes for peace, as God tells us to, in the midst of messing up?

If you find yourself needing to apologize, offer up an apology the right way. Many psychology experts agree that there are six important parts to a good apology. The next time either of you makes a hurtful mistake, whether it be unintentionally or intentionally, follow these six often-used steps to offer a sincere apology. Not all of them are necessary every time, but the more you can do, the better.

1. **Express regret.** Tell your spouse that you understand that you caused them pain. This shows that you see that you have hurt them and that you care about their feelings.
2. **Explain what happened.** Don't justify your actions; simply share an explanation.
3. **Take ownership.** Accept and acknowledge that you did something wrong. Saying "it's my fault" can go a long way.
4. **Repent.** Make a declaration to turn around from causing the same hurt down the road.
5. **Offer repair.** If there's something you can do to make it better, do it.
6. **Ask for forgiveness.** This is the final step in healing. Ask them to move forward with you.

Express Regret

Explain What Happened

Ask for Forgiveness

6 Essentials of a Good Apology

Take Ownership

Offer Repair

Repent

When hurt enters your marriage, and especially if it comes from your spouse, it's painful. But it will happen—we're human, after all. When it does, be prepared instead of alarmed, because we know with certainty that God desires connection and peace for your marriage, just as he desires connection and peace within the Church. Your marriage is his greatest example, so let's pursue it.

1. How difficult or easy is it for you to say "I'm sorry"? Has it always been that way?
2. Not all apologies are the same. How would you describe a wrong way to apologize? What does an unacceptable apology look like to each of you?

3. How would you rate each other's ability to apologize? How would you rate your own ability to apologize? Look over the six steps—which ones do you need to get better at?
4. What other Scriptures can you find on the benefits of apologizing and the dangers of not apologizing? Do a little research and share what you learn with each other.
5. Verbally demonstrate three different ways to say "I'm sorry." Practice saying it in a dismissive tone or disingenuous tone versus a genuine tone. Discuss the differences.

Dear Lord,

Thank you for being a God of grace, who so easily forgives us. May we respond in the same way to one another when we make mistakes that cause pain. Help us to grow in humility and in grace, so that we can have a marriage filled with your peace.

In Jesus' Name, Amen

I'll Be There for You

"A friend loves at all times, and a brother is
born for adversity" (PROVERBS 17:17).

WE IMAGINE that a budding friendship between the two of you likely formed a foundational part of your relationship early on, exactly as it should have. And before you were married, I'll bet your friends were really important to you. Hopefully, they still are. We all need people who know us, and our friends have an inside look into our lives. When we allow people to be close to us, they know what's going on in our hearts and minds as we live out our vows each day.

God may be the only one who fully knows you. However, as you grow in fully knowing your spouse, you should also sustain relationships with close, married friends, allowing them to know you as much as possible. There's a good reason for this. People who know you can take notice when something isn't right in your lives. Your friends can hold you accountable when you need attitude adjustments within your marriage, or if you could do with a little more humility, and they can use their objective perspective to look out for your marriage. We all need close friends to "love us at all times"—cheering us on in times when we're doing well and calling us out in times when we're struggling to be the best that we can be.

When we first met Matt and Karen, they were in a hard spot, but the only reason they decided to go to counseling was the encouragement of their best friends, who noticed they were hurting. In particular, Matt's friend knew him well enough to recognize his increasing stress levels and notice the ways Matt pulled away from his wife. Though it must have been a difficult conversation, his friend was willing to address it, because that's what real, Proverbs 17 friends do.

A real friend is faithful, willing to expose you for the sake of restoration. A friend who just tells you what you want to hear or ignores the difficult parts in your life is hardly a friend at all. Just like Matt, find and befriend people who will advocate for your marriage and love you at all times. Find friends who will grow with you, providing fun and laughter in your life and moving alongside you as a couple. When you find friends who share your beliefs and your desire for a God-centered marriage, you've found something really good. Securing friends who cheer you on and hold you to truth is one of the best things you can do for your marriage.

1. Are you satisfied with the number of friends you have as a couple? How many of them would you consider to be close friends? Name them together.
2. Is it easy for each of you to let people see your real life? Why or why not?
3. We often hear from people that they do not need friends. What are your thoughts on this?
4. What are some traits that each of you like the most in friends? Do you have these kinds of friends? If not, what is stopping you from making them?
5. Discuss and make a plan to spend time with another like-minded couple. Invite them over for dinner and a game night or offer another suggestion for a night out with them.

Dear Lord,

Thank you for teaching us what real friends are and how beneficial they are to our lives. Please help us to find these kinds of real friends— the ones who are likeminded and willing to speak truth. Provide us with opportunities to spend time with them and grow with them. May our marriage also be an example for our friends.

In Jesus' Name, Amen

Refined like Silver

"Husbands, love your wives, as Christ loved the church and gave himself up for her, that he might sanctify her, having cleansed her by the washing of water with the word" (EPHESIANS 5:25-26).

TO EXPLAIN HOW husbands should love their wives, Paul drew a parallel between the love of a husband for his wife and the love of Christ for his church. A husband should have a Christ-like, sacrificial love that elevates his wife and emphasizes her growth. Just as Jesus gave himself to transform us into his likeness, a husband should give himself for his wife, joining in this goal of sanctification, or helping her to become more like Jesus.

Like most growth, the process of sanctification is uncomfortable, requiring dedication, discipleship, and accountability. Zechariah 13:9 describes God's process as an intense fire, refining us into silver. Sanctification involves a small space and a lot of heat so that every impurity comes to the surface. Once exposed, each one can be dealt with and scraped off. Then, with every removed impurity, the silversmith adds more heat so any remaining impurities will begin to surface—and the process repeats. God the Refiner will know he's done when he can see his own reflection in the metal.

What a great image of sanctification to remember! God, the greatest silversmith, is working to reveal his image in your marriage and to see his own reflection in each of us. This is an amazing process, indeed, but it's not particularly fun to be under the heat. However, you are invited to take part in this sanctifying process together, as you take the time to deal with your impurities and help one another look more and more like Christ. You are called to love as Christ loves his Church—respectfully, sacrificially, joyfully, and continually. And this love transforms, as God works through this miraculous process meant for your good. Even though it may not always feel good to allow your "impurities" to come to the surface, you can trust that God has a great plan to use this process for your growth and for his glory. For in the end, your love for one another and your love for God will only increase, as his great transformative work is revealed.

1. How do you feel when one of your flaws comes to the surface? Have you ever thought about looking at them as an opportunity for growth?
2. Be honest with each other and share one "impurity" you see in yourself. Ask your spouse if they see that in you as well.
3. How do you currently handle it when your spouse's flaws come out?
4. Ask your spouse what you can do to help them grow spiritually and as a person right now.
5. Identify one area where you need some growth, then choose one thing you are willing to work on this week.

Dear Lord,

Thank you for not leaving us to stagnate where we are and for always desiring for us to grow. Help us to grow, change, and become more like you, and empower us to love one another in this refining process. Even though it may be uncomfortable, bring anything keeping us from knowing and demonstrating a greater love to the surface. Help us to reflect your image.

In Jesus' Name, Amen

You've Got a Friend in Me

"A man of many companions may come to ruin, but there is a friend who sticks closer than a brother" (PROVERBS 18:24).

WHEN WE FIRST GOT MARRIED, one of Chris's favorite things about our relationship was our friendship, and we've found this to be true for many husbands. Chris really enjoyed that I was someone he could have fun, dance around, and play with. To this day, our friendship is one of our greatest strengths as a couple. But we all know that friendship means so much more than just hanging out and having fun together, right? It also means midnight heart-to-hearts, shoulders to cry on, shared secrets, and truth-speaking. It means having someone who's willing to stick by your side, no matter what.

When Proverbs 18:24 uses the word *sticks*, it actually carries the same meaning as *cleaving* from Genesis 2:24—a man should "cleave to his wife." That kind of sticking, or cleaving, is permanent. It's like gluing something together so that it may never come apart again. And it's the kind of friendship you should strive for in your marriage—one that endures without disintegrating.

A lot can happen in your friendship that will strengthen your bond as a couple. For starters, laughter is linked to increased oxytocin, a chemical messenger in your brain that helps build trust and increases empathy. And who wouldn't want more trust and empathy in their marriage? Along with laughter, tears promote the release of oxytocin and endorphins, a built-in stress-reducer and pain-reliever for both emotional and physical pain. You can trust that those late-night, shoulder-crying sessions are just as important as those times of playing and laughing together. They both serve a greater purpose, making each of you a source of comfort and support for the other. They create a deeper need for each other and a stronger drive to really stick together.

Make sure you're making time in your marriage for friendship. Don't underestimate the powerful impact this investment can have in your marriage. There are so many simple ways to form a friendship together, so find what works for the both of you. Be the couple who laughs, plays, and leans on one another. You won't regret it.

1. Building a friendship takes time. What are you prioritizing over spending time together that maybe you shouldn't be? Is there anything you can let go of to allow more time together?
2. What are some things that you both find fun? How can you incorporate more of these into your marriage?
3. On a scale of 0 to 10 (with 0 being nothing and 10 being everything), how much of your real heart do you feel like you can share with your spouse? Why did you choose that number?
4. Who was your best friend growing up? What did you love most about them? What are the things you used to do together?
5. Make a plan over the next week to text each other a funny meme each day, or sit down each evening and share a joke you find humorous.

Dear Lord,

Thank you for the gift of friendship. Thank you for making it something that we can enjoy and also something that bonds us together. Give us more opportunities to grow our friendship, and help us to use our time wisely, so that we always have time for each other.

In Jesus' Name, Amen

For Better or for Worse

". . . he is like a man building a house, who dug deep and laid the foundation on the rock. And when the flood arose, the stream broke against that house and could not shake it, because it had been well built. But the one who hears and does not do them is like a man who built a house on the ground without a foundation. When the stream broke against it, immediately it fell, and the ruin of that house was great" (LUKE 6:48-49).

"Look, you want to know what marriage is really like? Fine. You wake up; she's there. You come home from work; she's there. You fall asleep; she's there. You eat dinner; she's there. You know? I mean, I know it sounds like a bad thing, but it's not." EVERYBODY LOVES RAYMOND

THERE'S A LOT WE COULD SAY about knowing your spouse will always be there. It's not something you have to think about or question; it's just an underlying, present trust. But coming to this security in your marriage doesn't happen naturally. It's earned. It's built when, time-after-time, things go wrong and your spouse still remains.

As much as we all hate the times of struggle, disconnect, and failure, there is a special gift hidden in their midst—the knowledge that, no matter what happens, your love is secure. Each difficulty is a chance for your love to say, "I am not going anywhere." Every trial you overcome together provides added security in your marriage. When we know and believe that love does bear all things, believe all things, hope all things, and endure all things, we can find security.

Do you know any couples who have seen their struggles through? Think about the admiration you had for them when chronic health issues, financial strains, or maybe even something as difficult as adultery hit their marriage, yet, they endured and actually came out stronger. Journeying through the hard times deepens the roots of your belief that, together, you can survive adversity and persevere.

Be sure to remind your spouse often that you're on this journey with them, and, even when the walls fall down around you, your foundation will stand firm. Don't allow times of trouble to derail you. Instead, use them to build a firm foundation of security and trust in your marriage, brick by brick. Finally, know that God is the one who builds this foundation, the only one who could make such an enduring love possible.

1. What would having security in your marriage mean to you? How would it make you feel?
2. Have you seen an unshakeable foundation in any couples you know who suffered hardships? Did you ever watch your parents go through trials? Discuss this together.
3. What do you think would be a difficult situation to endure in your marriage? What are one or two scenarios you are afraid of potentially encountering?
4. What are your thoughts on giving up when problems arise or when things get rocky? How do you normally handle turmoil or trauma?
5. Claim Luke 6:48–49 over one another. Make a promise to one another that you will not have a marriage built on sand that can be washed away. Write a note in your Bible next to this verse that simply says, "Our home." Use it as a constant reminder to stay the course when trials come.

Dear Lord,

Thank you for loving us with an all-encompassing, enduring love. Help us to love each other in the same way. Show us how to cling to you when we face hardships in our marriage. Take all that was meant for harm in our marriage and turn it into something good, and help us to trust you in the process.

In Jesus' Name, Amen

Steps in the Same Direction

*"Behold, how good and pleasant it is
when brothers dwell in unity!"* (PSALM 133:1).

THE IDEA OF WORKING TOGETHER is not a new one. In fact, there are quite a few idioms regarding the concept of unity, and you might be familiar with Abraham Lincoln's famous quote: "A house divided against itself cannot stand." While Lincoln's stance on unity changed the course of equality for the United States, we can trace his belief all the way back to Jesus: "Every kingdom divided against itself is laid waste, and no city or house divided against itself will stand" (Matthew 12:25). There's nothing that destroys a nation or a marriage more quickly than disunity.

Disunity, a state of being divided on significant issues, implies that we have different purposes or agendas, and are heading in different directions. By its very nature, disunity tears people apart. Perhaps this is why Scripture repeats itself on many occasions about remaining unified as believers. God knows that, once we start walking off on separate paths, we start to lose focus and crumble. Unity plays a crucial role in our lives as believers, as well as in our lives as married couples.

Divorces happen daily, with the most commonly cited reason for them as "irreconcilable differences." A less legal term? Disunity. Ultimately, disunity is the lack of ability to come together any longer once division has occurred. As a married couple, you must guard against this, or disunity will convince you that your disagreements are irreconcilable.

Once you start losing your focus to love the Lord your God with all your heart, mind, and soul, like Matthew 22:37 tells us to, you will start to drift apart. Be vigilant, and don't let the divide-and-conquer tactics of the Enemy have any room in your marriage. Remaining unified as a Christ-following couple means running the same agenda and sharing the same priorities to love God and each other. It doesn't mean you always have to agree, or that you will never argue or debate a point you don't see eye to eye on, but you do always have to move forward together, unified in direction, purpose, and love.

1. What has your focus been as a couple? Have you been seeking to satisfy your own agenda, or are you still prioritizing God's will?
2. As a spouse, how good do you think you are at looking out for both of your needs in your marriage? Do you find it hard *not* to look out for your own happiness first?
3. What is one area where you disagree? How could it cause you to drift apart, or how is it already causing you to drift apart? Is one of you willing to sacrifice something to reestablish unity?
4. If you have any divorced friends, what may have been the cause? How did disunity play a role?
5. Why do you think the Enemy uses a divide-and-conquer strategy? What do you think makes this strategy work so well for him?

Dear Lord,

Thank you for being a God of unity who directs us harmoniously. Help our marriage continue to stand by unifying our hearts. Keep us from being distracted by the schemes of the Enemy, as we fix our eyes on you. May we be a couple who seeks to stay on the same page instead of running our own agendas.

In Jesus' Name, Amen

He Is Worthy

"Therefore let us be grateful for receiving a kingdom
that cannot be shaken, and thus let us offer to God
acceptable worship, with reverence and awe, for our
God is a consuming fire" (HEBREWS 12:28-29).

JASON AND MEGAN were having a hard time worshipping God like they normally do—not because they felt God was unworthy of worship, but because things in their lives weren't going the way they wanted. When healing didn't come, interviews didn't turn into jobs, and paychecks didn't roll in, they struggled to worship. In light of all they temporarily lacked, turning to God did not feel sufficient to them. And we'd be lying if we said that we've never been there before. God is worthy of our praise at all times, even when we are struggling. When your human nature causes you to disagree with his plan, or when you don't get what you're after—those are the times you can actually offer up the best sacrifice of praise. You can worship him in every *even if* or *even when.*

As a married couple, I'm sure you've already found out that the domestic climate will not always be carefree and prosperous. There may be seasons when you beg God to help you just to make it through. As a couple, you need to spend those dark seasons in worship even more than you do the sweet ones. No matter how hard life shakes us, we have a God who invites us into an unshakeable kingdom, and that's enough.

Of course, we're not suggesting you give God lip service; praise without heart is the last thing God wants from you (Isaiah 29:13). He doesn't want you to worship only when your expectations are met. He wants your worship to withstand time, to weather difficulties, and to come from a place that, despite your circumstances, allows you to *know* he is good. But God can

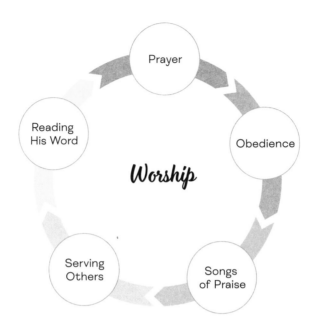

only receive your worship if you make an effort to give it to him; you have to participate.

As you go through this life together and enjoy the blessings he pours out, worship him. As you go through the unexpected hard times, worship him. And, as you go through the times when you don't even know if you can survive, worship him even more. No matter your circumstances, he will always be worth it—for all that he is, all that he's done, and all that he will continue to do.

1. Have you gone through a season together where it has been difficult to give God your worship? When was that and how did you arrive at that point?
2. Using the graph pictured, which area of worship do you think you need to improve upon? Which one do you feel competent in?

3. What are the ways you worship as a couple? What are some new ways you can worship together? Are you willing to put one into action this week?
4. How do you think serving others can be an aspect of worship in your lives? Can you think of one way to do this together?
5. What is your favorite worship song to sing to God? What about it do you find appealing?

Dear Lord,

You alone are the one God who is worthy of our praise and worship. You are the God who gives and takes away, and we want to have hearts that worship you in both. Thank you for all that you have already done for us and given us. Thank you for allowing us into your unshakeable kingdom. May our hearts of worship always be found acceptable to you.

In Jesus' Name, Amen

You're Not the Boss of Me

*". . . submitting to one another out of reverence
for Christ"* (EPHESIANS 5:21).

BEFORE WE CAN UNDERSTAND headship in the home, we must first understand that we all share equally the calling to submit, as followers of Christ.
In Ephesians, submitting to one another means fulfilling the call from God
to take on a servant's attitude—to not place ourselves above or over anyone,
but, instead, to serve and lift up others, as we love and obey Jesus. We are *all*
called to this type of submission.

In a marriage, specifically, a wife is called to submit to her husband
(Ephesians 5:22). Let's be clear—the husband is not called to demand her
submission. Submitting to a husband is a willful, chosen act of obedience
to the Lord, not a method of control for people. Even more, submission was
designed and modeled by God, as Jesus himself submitted to his Father (John
5:30). They were fully equal in deity, yet Jesus chose to serve and submit to
God's will.

Jesus' surrender had a purpose, and a wife's submission to her husband
does, as well. The wellbeing of a body thrives under good headship. The
members of the body are not less-than for trusting the head to take responsibility. Likewise, submitting to your husband should never mean submitting
to inequality, becoming a doormat, or assuming a position of less value.
Submission is part of the design for helping your marriage thrive. The husband maintaining headship in the home does not grant him control; it gives
him responsibility.

Remember what you model in a God-centered home—the submission to
the husband should look like the submission of the Church to Christ. This
means that a husband should take special care of his wife and family, and

that he will answer for how much his home exemplified the gospel of Christ. A husband should lead and love his wife so well that she would never even consider not following or trusting him. God calls husbands to a headship that values, loves, and honors the wives they lead. She is an equal, with an equally important role that adds value to society, the Church, her family, and her marriage. Ultimately, the leadership God desires is a leadership first submitted to him—one done with love and divine purpose.

1. What is your understanding of submission?
2. Do you have a positive understanding of submission, or does it feel like a negative, controlling act?
3. Read Ephesians 5:22–32. What does it specify about the calling of the husband and the wife? How does this alter or expand your understanding of submission?
4. As the husband, what does it mean to you to carry the responsibility of leading your home? Was this done well in your family? As the wife, how do you feel about being in a submissive role to your husband? What did this look like in your family?
5. Discuss what may have happened if Jesus did not willingly submit to his Father? In light of this, do you see submission as a positive act and state of mind?

Dear Lord,

Thank you, Jesus, for being willing to submit to the Father and to the cross, so we can have eternal life with you. We may not fully understand the plan of submission, but we trust you and your plan. Help us to surrender to each of our roles in marriage and to model Christ and his Church well.

In Jesus' Name, Amen

The Real Enemy

"For we do not wrestle against flesh and blood, but against
the rulers, against the authorities, against the cosmic
powers over this present darkness, against the spiritual
forces of evil in the heavenly places" (EPHESIANS 6:12).

JAMES AND SYDNEY, once happy newlyweds, found themselves at war within two years of marriage. They started having little squabbles, like many newlyweds do when they first start living together. However, these typical arguments quickly began to escalate, leaving them angry or silent for days at a time. Their fights over leaving the bathroom counter a mess or forgetting to clean up half-empty cups from the coffee table turned into personal attacks of being uptight or lazy. James and Sydney needed help, and they needed to learn to lay down their weapons of defensiveness and accusation. They needed to remember that their spouse was not their enemy.

One of the Enemy's greatest tactics is deception about his identity—distracting you from his attacks and leaving someone else to take the blame. He knows if you begin to fight your spouse and view them as the opposition, you free him up to wreak havoc in your lives. However, Scripture tells us clearly that your spouse never was and never will be the enemy. Why? Because your enemy is not flesh and blood. This deception is strictly a ploy of the Enemy to divide and conquer what God has joined together. Just like everything else he does, this is his attempt at corrupting with evil what God created for good.

As you move forward in your marriage, remember that your spouse is not your enemy, even if it sometimes feels like it. When James and Sydney learned to stop seeing one another as the enemy, but as allies who are on the same side, they could move forward safely. They no longer needed protection from each other, but from the Enemy and his attacks. Your spouse isn't seeking to cause you harm. In fact, your spouse likely wants what you want—a peaceful, happy marriage. The next time a little disagreement comes up, before letting it escalate, keep in mind who your real battle is against.

1. How do you think your real Enemy would benefit by turning you and your spouse against one another?
2. Do you have a plan for how you'd respond if your marriage was caught in a war zone, like James and Sydney's marriage?
3. Are there other people or things you're mistaking for the enemy right now? Can you recognize how Satan might be the real enemy in that situation?
4. Satan tends to cause doubt, make accusations, and bring shame. Are any of these personally affecting you or invading your marriage?
5. In what ways do you allow the Enemy to use you that might cause harm to your spouse?

Dear Lord,

Thank you for opening our eyes to see that the Enemy is our real opponent. Thank you for equipping us with all that we need to defeat him. Protect our marriage from his tactics, and help us to always work together to fight the real battle.

In Jesus' Name, Amen

Lead the Way

"Whoever walks with the wise becomes wise, but the companion of fools will suffer harm" (PROVERBS 13:20).

ON ANY GIVEN SUNDAY, when attending worship together, we'll see several elderly couples together when we scan the crowd. Many of them sit holding hands, leaning over and whispering to one another during the message, and exchanging quick glances and smirks as something the pastor says resonates with them. We sit in sweet awe of their undeniable love for one another. We enjoy where we are now, but we also long to have what these mature and experienced couples do.

We adored one couple in particular, Jack and Barbara. They have both passed away, now spending eternity with Jesus, but these two were a dear couple. As individuals, they were as different as night and day, but in their marriage, they shared a desire to follow Christ with their lives. Jack was a retired pastor, and Barbara was a feisty realtor. They spoke loads of wisdom into our marriage, as well as many other marriages behind them in years. They shared with us their hard days, their victories, and their insight from all the lessons they learned together. Some of them involved heavy failure and great humility; others involved extreme focus on sharing Jesus with everyone they met. They played an important role in our marriage, and we miss seeing their love on display.

Every couple needs a mentoring couple—their own Jack and Barbara. God's Word tells us there is nothing new under the sun (Ecclesiastes 1:9); whatever we face now or will face one day, someone else has already gone through it. When we have mentors, we reap the benefits of learning how to handle challenging situations, endure hard times, and celebrate all the wins. Mentors help teach us what matters and remind us that things will come to pass.

Time spent in wise mentorship always proves worthwhile and adds more value to your marriage than you can imagine, providing you with the clear vision you need to see around corners and the tools to face what's ahead. Beyond all their insight, they pour into you relationally, modeling perseverance and showing what it means to walk in grace and forgiveness. No mentoring couple has done or will do everything right, and that gives you the freedom to fail your way forward, too. Having mentors means always having a resource, a support system, and a team cheering you on. Let's be honest— who couldn't use that?

1. Do you have a mentor couple already? Discuss what benefits you have received from them. If you don't have a mentor couple, is there a couple you would like to mentor you now? Discuss this together.
2. What insight do you think you could glean from an older, wiser couple?
3. What questions would you have right now for an older, wiser couple?
4. What traits would you like to see in a couple who would mentor you?
5. Is there a couple behind you in marriage or pre-marriage that you could invest in and mentor yourselves? Discuss what that would look like.

Dear Lord,

Thank you for the gift of wisdom that you have given others and allow them to share. We don't want to be foolish in our marriage, and we ask that you provide wise couples, including a mentor couple, to walk alongside us. Allow us to be poured into, so we can also pour into others.

In Jesus' Name, Amen

In-Laws or Outlaws?

*"Honor your father and your mother, that your
days may be long in the land that the LORD
your God is giving you"* (Exodus 20:12).

WHEN YOU MARRY your spouse, you're marrying into their family. While your ultimate calling is to leave your parents and cleave to your spouse, family connections do matter to God (Genesis 2:24). But that doesn't mean they're always ideal or possible. You both may be blessed with tremendous parents, who pour into your marriage, share your faith, and encourage you to walk in truth as a couple. If that's the case, pause and praise God right now—really, he deserves a thanks for that kind of gift.

For the most part, couples struggle in the in-law department. Your new parents may be well-intended, but a little too intrusive, or they might be downright toxic. Whether your in-laws are thoughtfully overbearing or completely disruptive to your marriage, you can and should still honor them. Honoring them simply means you can treat them with respect, kindness, patience, and love, as God commands, regardless of how they treat you. However, you can achieve this while putting appropriate boundaries in place.

Before your new extended family, your ultimate priority lies with your new immediate family. If anyone, including either set of parents, causes them pain or disrupts their well-being, you need to address it. Your first goal is to keep the peace, as much as it is up to you; in other words, do *your* part (Romans 12:18). You are not responsible for your in-laws' behavior, but you *are* responsible for helping them understand what's acceptable to you and what isn't.

Not all in-laws react to new spouses, or any kind of change in the family, in healthy ways. If this is the case in your marriage, place appropriate boundaries with love and kindness, and, if you need it, seek outside help. Ultimately, the spouse whose parents are causing the struggle has the responsibility of addressing it. In-law struggles don't have to have a negative impact on your marriage. You can deal with in-laws with compassion and know that positive outcomes do happen.

1. What are your relationships like with both sets of your parents? Are there any changes you'd like to see?
2. Do you feel like your spouse has your back when it comes to your relationship with their parents?
3. Are there things you love and appreciate about your in-laws or your relationship with them?
4. Are there any boundaries you would like to put in place in order to protect your family?
5. Is there a need for outside help with your in-law situation? Spend some time looking up a how-to for setting boundaries with in-laws. If this is a problem in your marriage, make a plan today.

Dear Lord,

Thank you for equipping us to love and honor our parents, even when it's hard. Help us to learn to appreciate all that we can from them, while remaining loyal to and protective of one another. Together, may we show grace, honor, and kindness, as we seek to do our part in all of our family relationships.

In Jesus' Name, Amen

More Than a Feeling

"Love is patient and kind; love does not envy or boast;
it is not arrogant or rude. It does not insist on its own
way; it is not irritable or resentful; it does not rejoice
at wrongdoing, but rejoices with the truth. Love
bears all things, believes all things, hopes all things,
endures all things" (1 CORINTHIANS 13:4–7).

MISCONCEPTIONS ABOUT LOVE run rampant in our culture, both inside and outside of the church. We see so many couples who lack an understanding of what love really is. When Nick and Alana came to see us, counseling was their last-ditch attempt to make their marriage work. Nick told Alana he had "fallen out of love" with her. He reluctantly came into our office, already believing the marriage was over. Alana, heartbroken and desperate, wanted to understand what happened to change the love Nick had when he first said "I do."

In Nick's mind, and in our cultural context, love is a hole in the ground—something you can fall into and stumble back out of. This belief underlies many divorces. It is shallow, feelings-based, and passive. It depends upon your spouse fulfilling all of your needs, constantly digging to keep you falling deeper. This is a spouse-centered love, not God-centered, and it's unsustainable.

God makes the love he calls us to very clear throughout Scripture. His love is sacrificial, patient, and enduring. His love holds on when the going gets tough, instead of letting go and falling out. His love endures, even as feelings rise and crash tempestuously, like waves in the ocean. The emotional side of love is important, however, it's just one element among the depths of love God desires us to have for one another. Should we throw an entire marriage away because one aspect of it is missing or deficient? Should we settle for "falling out of love" when we can plunge boldly into his real love in our marriage? Of course not.

The love God desires for your marriage is as easy as it is hard. It grows through perseverance and endurance, and lacking emotion for your partner should not be allowed to weaken it. This is an issue that can and should be addressed. When Nick began to understand what real love called him to, he was willing to give it one more shot. Together, they began to repair the injuries that had caused their love to dwindle in the first place. Now, Nick and Alana have a marriage they know will endure all things—all because they chose to seek God's definition of love instead of their own.

The Love of God
(1 Corinthians 13:4-7)

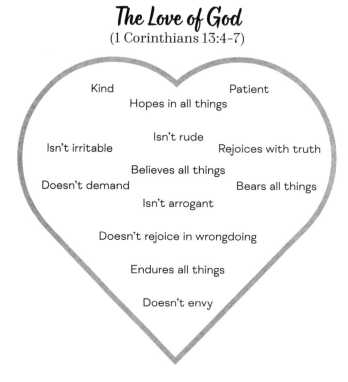

Kind
Patient
Hopes in all things

Isn't rude
Isn't irritable
Rejoices with truth

Believes all things
Doesn't demand
Bears all things

Isn't arrogant

Doesn't rejoice in wrongdoing

Endures all things

Doesn't envy

1. In your own words, what do you say love is?
2. What did your parents and their marriage teach you about love?
3. What are some things you can do in your marriage that will help the emotional part of love stay alive?
4. Using the 1 Corinthians love chart, go through each aspect of love and discuss with one another how you would rate yourself at displaying those traits.
5. Which aspect(s) of love do you currently need most from your spouse? Which could you improve upon?

Dear Lord,

Thank you for providing us with the real definition of love—a definition that allows us to remain together and endure whatever life throws our way. Thank you for Jesus, who showed us this type of love in the most sacrificial way possible. Help us to display your love in our marriage.

In Jesus' Name, Amen

Anxious for a Reason

"The unmarried man is anxious about the things of the Lord, how to please the Lord. But the married man is anxious about worldly things, how to please his wife, and his interests are divided" (1 CORINTHIANS 7:32-34).

DID YOU KNOW that when you said your vows, you also signed up for a life of anxiety? Yeah—we didn't either. Despite how awful or discouraging that may sound, it's really not as rough as it seems. You will experience a certain amount of stress, but it doesn't have to hold you captive—you can cast your cares upon the Lord and exchange them for his peace (1 Peter 5:7).

It's important to remember where there is anxiety, there is love. We're told upfront we will have anxiety in our marriage, because God knows if we love our spouse, they will become a competing object of our attention. The unmarried person only needs to please the Lord; the married person will now be tempted with trying to please the Lord *and* their spouse. Desiring to please your spouse isn't a bad thing, but when we lose focus of God's call to seek him first, we can easily drift from a God-centered marriage to one that is spouse-centered.

You probably know at least one couple where one spouse is always in charge and the other walks on eggshells to please them. This was the case for Cherise and Michael. Cherise always wanted to go to church and have a family-devotion time, but Michael wanted to enjoy sleeping in and constantly seemed too busy to spend purposeful time with the family. Not wanting to upset Michael, it didn't take long for Cherise to fixate on pleasing her husband instead of pleasing God, and she gave up going to corporate worship and leading her children in his Word. As a result, Cherise began to fall away from the Lord personally, as she slowly disconnected from her church family and stopped reading the Bible to her children. This constant battle between what God wanted and what Michael wanted was filled with tension.

Understand that this battle will take place in your marriage. Sometimes, pleasing the Lord will upset your spouse. You are not powerful enough to always please your spouse, so this cannot be your main objective. As a couple, seek God above all else. When you do, he will remain at the center of your marriage, keeping this competitive anxiety at bay.

1. What areas in your marriage do you worry about making your spouse happy? Explain.
2. In your marriage, where is there tension between doing what God wants you to do and what your spouse wants?
3. How does this affect your relationship with your spouse? With God?
4. Look up the definition of the word *anxiety*. How can anxiety be harmful in your life?
5. What you treasure is often where your anxiety lies. What are the things in your life that you stress over? What might they suggest you treasure?

Dear Lord,

Thank you for being a God who is jealous for our love and worthy of our affection. Thank you for also giving us each a spouse who we love and desire to please so strongly. Help us to balance out our love for you and for them, while we keep pleasing you as our highest priority in our marriage.

In Jesus' Name, Amen

Is This Working?

"Whatever you do, work heartily, as for the Lord and not for men, knowing that from the Lord you will receive the inheritance as your reward. You are serving the Lord Christ" (COLOSSIANS 3:23-24).

THE WORK STRUGGLE is real, and often affects husbands and wives differently. Early in his marriage, Alex struggled with his career choice. He wanted to do what he loved but also make enough to provide for his family. More often than not, career opportunities that paid enough were not the ones he really desired. He had a dilemma—choose personal fulfillment or better financial provisions. Meanwhile, Kelly faced her own choice—to be a stay-at-home mom or a working mom. Before they welcomed their first child, Kelly had a fulfilling job at a large bank, with several employees under her, but now she felt pressure from family to stay home with the baby. When they came to our office, unsure of the right choice, we encouraged them to turn to God's Word for help.

In Genesis 3, God tells Adam because of his sin, he will have to labor painfully—in other words, his work will be physically hard. Does this mean Alex should give up his dreams of one day having a career he loves? Of course not. It simply means that work will always require, well, work. For Alex and his family, nothing will come for free, and he can't evade this struggle altogether. Sacrificing the financial security of his family to do a job he loves isn't a viable option right now; for the time being, he is called to provide for his household (1 Timothy 5:8).

However, with the Colossians verse above in mind, Alex can find fulfillment no matter the job, knowing God is his reward. If his dream job comes along with his dream paycheck, he can responsibly and joyfully accept it. Still, he can use his gifts and pursue his desires in his current roles, inside and outside of work. By taking on a servant attitude and working for the Lord, Alex has the freedom to enjoy any job he chooses—as does Kelly. Make no mistake, the Proverbs 31 woman does take care of her family, but she also makes and sells garments, purchases land, plants a vineyard "out of her earnings," turns a profit, and "sets about her work vigorously." As a wife and mother, you get to decide to stay home, pursue a career, or do both throughout your life. The choice belongs to you. As a couple, you can decide together for your family.

God has granted every single person a variety of gifts and strengths, and fulfilling the purpose God created you for matters most. We won't tell you what Alex and Kelly chose in the end, because it really doesn't matter. The important part is that they discussed it together, decided what worked best for their family, and what aligned most with God's will for them.

1. Assess your current work situation . Are the family's financial and emotional needs being met? Is God being honored?
2. How have you struggled with job decisions? What factors are important to you as a couple and to God?
3. If you could choose any career, what would it be? What are some things you could do today to make it happen later, when the time is right? Would it be okay if you never got to do it?
4. How is it possible to find fulfillment in a job that fits your family life, even if you don't really like it?
5. Is it ever appropriate to quit a job? What cost is too high for your family to pay when it comes to unhappiness at a job?

Dear Lord,

Thank you that we can find delight in any career decision we make, even though it may not be the perfect path we dreamed of. Thank you for giving us each distinct desires and gifts, and for creating us uniquely to work in different capacities. May our career choices be honoring to you and serve a purpose higher than we could ever imagine.

In Jesus' Name, Amen

The Value of Things

"But godliness with contentment is great gain, for we brought nothing into the world, and we cannot take anything out of the world" (1 TIMOTHY 6:6-7).

FOR YEARS, couples have been fighting about money. When handled with care, your money can be a wonderful resource for blessing your marriage and serving others. It can help further the kingdom of God. So it's no surprise that the Enemy would seek to distort and exploit your finances as a weapon against your marriage. Read a little further into 1 Timothy, and you'll see that "the love of money is a root of all kinds of evils" (1 Timothy 6:10).

What makes dealing with money so tricky in your marriage? Long before you met each other, you already developed an idea about money and what it represents on your own, with unique upbringings that informed your financial habits. Money can mean very different things to different people, so when money struggles come up, these differing beliefs come up with them. It's important to talk openly about your financial beliefs together, but it's also important to know what God says about money:

- God is the one who meets our needs, not money (Philippians 4:19).
- With or without money or possessions, we must be content (1 Timothy 6:6-8).
- Integrity is more important than wealth (Proverbs 19:1).
- We are to give generously and to give in faith (2 Corinthians 9:6-8).
- Your spending habits will reveal what you treasure (Matthew 6:21).
- We are to tithe and give God our first fruits (Malachi 3:10, Proverbs 3:9-10).

If you want to avoid the dreaded fights over finances, get on the same page about your money now. Having a monthly budget, a spending plan, a plan to tithe, and agreed-upon financial goals are foundational pieces for living stress-free financially. It's never too early or too late to get your finances in order. You can change poor spending habits and pay off debt when you have a plan. As you navigate finances in your marriage, the only thing you really can't afford is denial.

What Money Means to Me	What God Says
Status	Everyone is equal. GALATIANS 3:28
Pleasure	Fullness of Joy is found in God. PSALM 16:11
Control	God is in control. PROVERBS 19:21
Security	We are secure in God. JOHN 10:29
Gifts	Our treasures are in heaven. MATTHEW 6:20

1. What was the financial state of the home you were raised in? What were you taught about money, and how did that shape your beliefs?
2. Denial and myths about debt are common. What are your current beliefs about debt?
3. Do you wish your debt ratio or financial habits were different than they are right now? Discuss your goals and ideals.

4. Take a look at the left side of the chart and choose your top two answers of what money means to you. Discuss your answers and compare them with what God says.
5. If you haven't already made one, make a financial plan. Discuss what this process looks like (i.e., taking a course, reading a book, making a budget sheet, and so forth). Set a date to discuss your plan, and stick to it.

Dear Lord,

Thank you for the gift of money and the freedom to use it to enrich our lives and further your kingdom. Help us to find contentment in our seasons of plenty and our seasons of little. Help us to keep our focus on the things that are truly essential, and allow our spending habits to reflect how much we treasure eternal pleasures over temporary ones.

In Jesus' Name, Amen

More Than Multiplying

*"Let your fountain be blessed, and rejoice in the wife
of your youth, a lovely deer, a graceful doe. Let her
breasts fill you at all times with delight; be intoxicated
always in her love"* (PROVERBS 5:18-19).

BE HONEST. How comfortable did you feel reading that verse together? It's okay—conversations about sex and intimacy can get a little awkward. But, between a husband and wife, they shouldn't be. In the Christian world, it's not uncommon to hear that God created sex for you to "be fruitful and multiply" (Genesis 1:28). However, Christian sex talk tends to end there. It rarely sounds like the tone God portrays in Proverbs 5. Not all Christians know that sex was *created* for so much more than having children. After all, what would this limited "be-fruitful-and-multiply" ideology reduce sex to for married couples who can't conceive or who decide not to?

God is a God of intimacy. Think about how much he loves relationships and closeness; he sent his Son to die, so we could be in a relationship with him. When he first made Adam and Eve, he made them naked and unashamed—fully, visually, and relationally intimate with one another. He knows every detail of our humanity and our stories, from knitting us together in our mother's womb, until the end of our lives (Psalm 139). How much more personal can you get than our God and his design?

In its truest form, sex facilitates two becoming one in marriage—not only two physical bodies, but two hearts, two minds, and two souls. Sex provides physical pleasure, strengthens intimacy, and reveals your longing for devoted companionship. God desires more than procreation for your marriage; he wants you to find pleasure and delight with each other. We were made with a longing for connection, body and soul, and your sexuality cannot be separated from your spiritual identity. In the same way, sex and intimacy should not be separated from your marriage.

God intended marriage to reflect Christ's love for the Church, and he is far too creative and generous to limit the beauty of this union. A sexually satisfying marriage embodies and honors God's design.

1. As a couple, how comfortable are you talking about sex with one another? Were you taught to treat sex as taboo? What were you taught about the role of sex in a relationship between a husband and wife?
2. Do any of the God-designed benefits of sex surprise you? Which ones, and why?
3. Consider how you would you rate your sex life. Openly and honestly talk about ways to enhance this area in your relationship.
4. Many couples struggle with sex, especially if they've been taught incorrectly how to view it. How have you experienced struggles with sex in your marriage as it pertains to your view of it? What resources can you use to educate yourself and inform?
5. It's not uncommon for couples to experience physical sexual struggles—from navigating certain medications, pregnancy, hormonal imbalances, impotency, or aging. Make a plan for how you will deal with sexual problems that enter into your marriage. Decide now not to let shame have a voice in your sex life.

Dear Lord,

Thank you for the beautiful design of sex within our marriage. Thank you for giving it a purpose beyond procreation, so that we are able to enjoy closeness, pleasure, and greater intimacy with one another. Help us to always be open and honest with one another and to never lose the ability to physically connect.

In Jesus' Name, Amen

Can't Stop, Won't Stop

"In all circumstances take up the shield of faith, with which you can extinguish all the flaming darts of the evil one; and take the helmet of salvation, and the sword of the Spirit, which is the word of God, praying at all times in the Spirit, with all prayer and supplication" (EPHESIANS 6:16-18).

IF WE HAD TO NAME the most important component for your marriage, it would be prayer. Prayer gives us a direct link to the almighty God. It's like having access to the top expert in any field, task, or challenge. He's the best guide for any journey and the best protector against any enemy. As followers of Jesus, prayer gives us an incredible advantage in life and in marriage, and it's also a beautiful gift we can give to God and share with one another.

The only thing you need in order to pray sincerely is a spirit of helplessness—a soul that can do nothing apart from God. Fortunately for us, we have that. And praying doesn't have to be complicated to be incredibly powerful. You don't need eloquent words, extensive knowledge, a certain feeling, or the intervention of a priest—you only need a heart that seeks God.

When you pray, you tap directly into God's heart. You enter into the presence of the Father, who longs to sit with you, guide you, hear from you, and speak to you. He often longs for you to ask for his help, just so he can provide (James 4:2-3). He cherishes the moments we seek him earnestly, especially when we do it together. His presence is guaranteed, and he hears us when we ask for his will in our lives (1 John 5:14-15).

Your marriage will be most fulfilling with God at the center—and prayer invites him there. When you pray for one another, you ask God to go before your spouse in protection and provision. There is no greater act of love for your spouse than to bring them before the throne of God. Keeping your marriage surrendered to God's will and his plan is the best position to be in to receive all he has for you as a couple. Nothing will unleash God's will in your lives more than prayer, and nothing will serve your marriage more than kneeling together at his feet.

1. How do you feel about your current individual prayer life? Your prayer life as a couple? What would you change about it?
2. What struggles do you have with prayer? Are you comfortable praying together with your spouse?
3. What needs or desires do you have as a couple that you have not asked God for? Are you willing to pause and ask him to meet them now?
4. Did you have someone praying for you when you grew up? Who was it, and how do you think their prayers impacted your life? Who can you be praying for to impact their life?
5. Share your current prayer requests with one another. Tell your spouse at least one thing you would like them to pray for you regarding your spiritual growth. What's one thing you want to pray for in your marriage?

Dear Lord,

Thank you, Father, for desiring a relationship with us through prayer. Thank you that we always have the opportunity to come before you, night or day—that our unceasing prayers will always be heard. Give us the desire to long for your presence in our lives, so that we can remember to ask you daily to remain in the center.

In Jesus' Name, Amen

Three's a Crowd

"Let marriage be held in honor among all, and let the marriage bed be undefiled, for God will judge the sexually immoral and adulterous" (HEBREWS 13:4).

WE HAVE SOME BAD NEWS—no marriage is immune to the risk of adultery. And that risk isn't going away anytime soon. In fact, with social media and dating apps, having an affair has become even more accessible. This is why God instructs us to pray against temptation. He knows that even when our spirit is willing to keep our vows, our flesh is often weak (Mark 14:38). We live in a time where we no longer need to search out sexual temptations; they find us in our inboxes, televisions, phones, and other devices. Standing on guard has never been more important.

For years, men have had a higher rate of adultery than women, but this gap is closing. Google the latest statistics to track these numbers. With recent findings, neither of you should feel immune. You both have to be on guard, understanding that affairs are rarely just about sex.

After cheating on his wife, Hunter shared with us that the "other woman" had become his greatest supporter and encourager, especially since his wife, Sarah, seemed to point out his flaws constantly. Disconnected and uninvested couples are the most vulnerable. This is why friends and co-workers are likely the ones who lured in the cheating spouse, intentionally or unintentionally. In these cases, they were not drawn in by sexual desire alone but by affection and attention from an outsider. God gave us all a desire for love and acceptance, but, if given the opportunity, the Enemy will work to corrupt this with evil. He will convince you that an affair can satisfy this desire more than God can in your marriage. Meanwhile, affairs cheapen our God-given longing for acceptance and betray the ones we promised our love to.

Protecting your marriage against adultery will always involve prayer, open communication, support, and validating one another. If any one of these areas fall to the wayside, address it. When you fail to invest in your marriage, you also fail to protect it. Acknowledging that you both have a responsibility to maintain your vows is vital. There will never be a valid excuse or justification for an affair, for no one falls victim to committing adultery—it is always a choice, and it is up to you as individuals to walk away.

1. Is adultery something you are afraid of in your marriage? Why or why not?
2. What are some things you can do to prevent the temptation of adultery in your marriage? Is there a person in either of your lives who is or might become a temptation?
3. Have you ever witnessed how adultery affected another couple? What was the result in their marriage?
4. Do you believe that adultery is forgivable, or does it automatically end a marriage? What does God's Word say about it?
5. If your spouse ever did have an affair, tell them now what you think it would do to you. Out loud, finish this sentence: "If you ever cheated on me, it would make me feel like [. . .]." Take turns sharing with one another from your heart.

Dear Lord,

Thank you for giving us our marriage and creating my spouse just for me. Thank you for giving us the gift of one another—a gift that no one else is entitled to have. Help us to keep our marriage pure, our eyes from looking elsewhere, and our minds from finding affection in another person. We pray now against this temptation in our marriage.

In Jesus' Name, Amen

Can't We Just Get Along?

"Count it all joy, my brothers, when you meet trials
of various kinds, for you know that the testing of your
faith produces steadfastness. And let steadfastness
have its full effect, that you may be perfect and
complete, lacking in nothing" (JAMES 1:2–4).

MOST OF US don't enjoy conflict. But if you ever want to grow in your marriage, conflict is necessary. Those draining debates and disagreements will test your faith and produce steadfastness. Without the chance to work out your faith, it will remain weak and wavering. Likewise, without conflict in your marriage, there is no growth.

Trust us on this one—though it may seem like a bad thing in your marriage, conflict is really useful and productive. It allows you to each have a voice, to stand up for the things you believe in, and to learn to listen to and understand one another. Conflict itself is not unhealthy, but *how* you experience or react to conflict might be. Dr. John Gottman at *The Gottman Institute* has defined four of the unhealthiest behaviors that can present themselves during a conflict:

- Criticism: expressing disapproval of your spouse or the way they think, feel, or act
- Contempt: showing disregard or dismissive behaviors (i.e., eye rolling, showing disgust)
- Defensiveness: immediate excusing, justifying, and self-protecting our own words or actions
- Stonewalling: cutting off or refusing to communicate with your spouse (i.e., giving our spouse the "cold shoulder")

When you find yourselves in a disagreement, try to avoid anything that will cause harm and breed disconnection. Constantly criticizing your spouse places them in a defensive position, and even your valid opinions or complaints will go unheard. The same goes for contempt—if you disregard, mock, or use passive-aggressive tactics, your spouse will feel rejected. Degrading your spouse should never be the goal of any conflict.

Both criticism and contempt are a set-up for defensiveness, but neither of you have to take the bait. If you focus on your response and learn not to react defensively, this will allow your spouse to feel heard. When you listen to seek understanding instead of a better defense, you will draw closer together instead of further apart.

Lastly, remain engaged, and refuse to allow the silence of stonewalling in your relationship. Stonewalling automatically shuts down connection; there's no benefit at all in this for your marriage. To engage in conflict well, there must be respect, listening, and a desire to understand one another. Ultimately, you are each responsible for your own role in marriage conflicts, and God calls you as individuals to seek peace and reconciliation at all times.

1. Out of the four unhealthy ways to have conflict, which do you struggle with? Explain.
2. Ask your spouse if they see any of the four unhealthy forms of conflict in you. Be willing to listen, and lay down the desire to defend yourself. This is information you want to have.
3. Is there something you would like your spouse to do differently during times of conflict (i.e., change their tone, quit walking away, remain calmer, and so forth)?
4. How did your parents handle conflict in their marriage? Do you see any of their traits showing up in your own marriage? Discuss.
5. How do you think conflict can actually benefit your marriage and help you grow together? How can a lack of conflict be bad for your marriage? Discuss James 1:2-4, and also verse 19.

Dear Lord,

Thank you for making us with unique personalities that produce a variety of differing thoughts and opinions. Thank you for giving us the ability to use these differences so we can grow individually. Help us to be humble and in maintaining an openness to learn from one another.

In Jesus' Name, Amen

Do You Remember That?

"And Joshua said to them, 'Pass on before the ark of the LORD your God into the midst of the Jordan, and take up each of you a stone upon his shoulder, according to the number of the tribes of the people of Israel, that this may be a sign among you. When your children ask in time to come, "What do those stones mean to you?" then you shall tell them that the waters of the Jordan were cut off before the ark of the covenant of the LORD. When it passed over the Jordan, the waters of the Jordan were cut off. So these stones shall be to the people of Israel a memorial forever," (JOSHUA 4:5-7).

ON THEIR ANNIVERSARY, our oldest daughter and son-in-law, Alicia and Chad, plan a time of remembering. They sit down together to process all the year has brought them. They reminisce over their biggest moments and life changes, as well as any vacations they have taken or adventures they have embarked on. But their main focus is on remembering God's faithfulness in their marriage over the preceding twelve months. On one occasion, they recalled a financial blessing from an anonymous friend who paid their giant water bill from a busted pipe. Another year, they thought of how a major flood in their home turned into a remodeling project they had longed to do. They write down these memories of God's faithfulness and place it inside a little jar, along with a symbolic memento of the year, displaying it on a shelf in their home.

One jar has a receipt of the paid bill, another holds water to remind them of the flood. Even when your year holds pain or disappointment, or when what's inside your jar isn't what you expected, put them on display in your marriage. Don't treat these moments with less value or attention. Your memories don't need to be uncomplicated or social-media worthy. Seek to remember the moments that only you understand in your hearts. When you do, look for his faithfulness.

That is exactly what the Lord led Joshua to do. When God brought the Israelites into the promised land, he wanted them to remember how he dried up the Jordan and made the way for them. He wanted to remind them that he journeyed with them in all they went through. As they looked at those stones, they would see God's faithfulness in their story, and tell of it in times to come.

Remember God in your marriage. As the years fly by, take pictures, fill jars, and pause to recall all the Lord has done. May his faithfulness always be your greatest memory.

1. When was the last time you reflected on what God has done for you? Discuss a time when God proved himself faithful when you weren't sure he would be.
2. What do you think the purpose is in remembering God's faithfulness?
3. Reflect on the beginning of your relationship to wherever you are now. What has God already led you through, as individuals and as a couple?
4. Is there something tangible, like Joshua's stones or Chad and Alicia's jars, that you could keep or create to remember God's goodness together? How will you display these items?
5. What is your fondest memory of God's goodness growing up? Share your stories with one another.

Dear Lord,

Thank you for giving us tangible ways to look back and remember your goodness and the many ways in which you have provided for us. Thank you for drying up the rushing rivers of our lives, so we could make them into a life with your promises. Help us to never lose sight of your hand in our lives.

In Jesus' Name, Amen

Sundays Are Not for Sleeping In

"And let us consider how to stir up one another to love and good works, not neglecting to meet together, as is the habit of some, but encouraging one another, and all the more as you see the Day drawing near" (HEBREWS 10:24-25).

THE BED can seem extra comfortable on a Sunday morning. We get it. Oftentimes, the effort it takes to get up on Sunday reflects the effort we spent during our week. Many of us push and push until the great Sunday collapse comes along. This was definitely the case for one couple we worked with, Ethan and Rachel. After a full week of work and a Saturday spent on the ball-field with their son or catching up on chores and yard work, they were tired, and Sundays were tough. In their fatigue, they made the mistake of exchanging church for a morning off. Sleeping in a little longer on Sundays seemed like a good idea for the tired couple.

They needed rest. However, Ethan and Rachel chose a couple extra hours of tossing and turning in exchange for a morning of connection, worship, and encouragement to face the week ahead. They told themselves that church could be anywhere; after all, the Church is not a building, but God's people. But that's exactly the point. Sleeping in, going for a peaceful walk, and watching a sermon online are all okay on occasion, but getting into the habit of only doing these things causes you to miss out on the people gathering together as the Church. God cares deeply about his children taking care of one another, sharing each other's burdens, and rejoicing in each other's wins. He desires for us to worship together, to lift one another up in prayer, and to stir one another up in doing good. He delights in the sound of a mighty chorus, lifting up the praise that he so deserves.

The next time you're tempted to trade in meeting with God's people for a little extra sleep, remember the rest you're giving up—encouragement, refueling, and strength as the body of Christ in the local church. And maybe take a nap when you come home instead.

1. How do you currently feel about the church you belong to? Is it a healthy church for you as a couple?
2. Think about your involvement in your church. Are you as involved as you'd like to be? Are you over-involved, to the point where you're spiritually exhausted? Discuss this as a couple.
3. In what ways does your church help each of you to grow spiritually? How has your church encouraged you?
4. What are other things about going to church that might be important to God? What are other benefits of attending church together?
5. What are your friends like at your church? Do you have any? Would you like more? What plan of action can you make to grow your church community?

Dear Lord,

Thank you for giving us a faith community and the ability to meet together with your children. Thank you for the work of the Church, which is designed to teach, spiritually nourish, and connect us as your body. Help us to schedule our weeks in a healthy way, so we long for church on Sunday mornings instead of feeling too tired for it.

In Jesus' Name, Amen

First Things First

"But seek first the kingdom of God and his righteousness,
and all these things will be added to you" (MATTHEW 6:33).

DID YOU KNOW that the word "priority" was originally only used in singular form? It wasn't until the 1940s that the plural version came into play. At least in language, people began their day with one priority, one focus, and, when it was achieved, then they would move forward. Today, most of us live in a culture of hustle and grind. It wouldn't be strange if you looked at your calendar or to-do list right now and found a list of several *priorities* that demand to be done.

For a lot of couples, their minds start racing with all they need to do as soon as they lift their heads from their pillows. This can be true of husbands and wives in all facets of life. Regardless of what's on your priority list, it gets harder and harder to focus on just one. But, in God's goodness, we were given an example.

If we followed a day in the life of Jesus, as he lived on earth, we would see him do the same thing over and over—rise and seek his Father. I doubt that one day passed by where he did not give his Father full attention, first thing. He made time with God his priority. There was no close second—no plural version. Seeking God was his first step in putting the kingdom of God and his righteousness on the to-do list. As Jesus walked as a man, engaging with God in every moment and prioritizing the things that were important to his Father, it allowed him to accomplish the work of the gospel.

Even in sharing his divine status, Jesus made time with God the Father a priority. If this was Jesus' plan, how much more should it be our plan? When you rise in the morning, you can choose to go about your business your way by hitting the snooze button a few times, or catching up on social media with any extra time, or you can go about God's business, by making the time to pray together and asking the Lord what he wants from each of you for that day. Trust us—one of these is going to work out better than the other. God's Word holds true; when you learn to put first things first, God adds to your life the rest of what you need.

1. In thinking about your schedule or to-do list, what emotions does it evoke? Stress? Peace? Discuss your current workload and where you would like it to be.
2. Do you have a plan for the way you prioritize your day? How do things make it on your priority list? Are they all priorities?
3. What would your day look like if you only had one priority—seeking God first? What would your schedule and life look like if he was in charge of it?
4. How do you know when your lives are out of balance or overwhelming? How do you feel and act when this happens?
5. Share with one another your top priority for tomorrow. Then, share your to-do list for the week. See if there is a way you can support each other in prioritizing your time and schedule in healthy ways.

Dear Lord,

Thank you for being a God who is worthy of coming first in our lives. Thank you for having a purpose for each of us and a plan for how we can prioritize our time and resources. Help us to honor you and give you first place in our lives.

In Jesus' Name, Amen

The Languages of Love

*"Beloved, let us love one another, for love is
from God, and whoever loves has been born
of God and knows God"* (1 JOHN 4:7).

AS NEWLYWEDS, Isaac and Hannah were off to a great start in their marriage. They went through premarital counseling, read several inspiring and informative books together, and even invested in learning about each other's love languages, using the assessment from Dr. Gary Chapman's *The 5 Love Languages.* Each of us have preferential ways of receiving love—physical touch, words of affirmation, quality time, gifts, or acts of service. For Isaac and Hannah, Isaac receives love most through words of affirmation, and Hannah receives love most through acts of service.

God calls us to love one another, but we sometimes focus more on receiving love from one another than giving it. Isaac and Hannah loved each other well, but as the years went by, their family grew and their lives got busier. The more depleted they got, the more they focused on hearing their own love language instead of speaking their spouse's.

In their increasingly busy household, Hannah wondered why Isaac wasn't helping out more. Work left her exhausted, and the tasks at home kept piling up. She did what she could, like packing lunches and making the coffee every morning, but she just wanted Isaac to notice the overflowing laundry or the sink full of dishes. Meanwhile, Isaac wondered why Hannah didn't acknowledge him. He constantly told Hannah how much he appreciated her, but it never seemed to matter. Even when he worked his hardest, she never stopped to thank him or tell him he was doing a good job.

Can you see the cycle? Isaac and Hannah were falling into a trap many couples fall into. They wanted their spouse to fulfill them more than they

How Do You Receive Love?

Physical Touch

Words of Affirmation

Quality Time

Gifts

Acts of Service

wanted to love their spouse, as God calls us to. Of course, they didn't stop loving each other; they simply struggled to hear it, because it didn't come in the language they best receive it in. Hannah's primary love language—acts of service—left her to receive only love when Isaac did something for her. She was missing all the love he showed with his words. Isaac's primary love language—words of affirmation—also reduced his perception of Hannah's love, which she showed daily by packing his lunch and making the morning coffee. As a couple, you'll both have specific ways you appreciate being loved and natural ways of showing love. Remember that God's love speaks every language, so consider how to best show this love to your spouse and enjoy embracing all the ways they show love to you.

1. Do you know one another's preferred love language? Go to www.5lovelanguages.com to take the assessment or use the graph. Evaluate how well you and your spouse agree with your individual results.
2. God calls us to love one another. How do you naturally show your spouse you love them? What are some specific examples of what this looks like to you?
3. We all feel loved in different ways. Discuss your top two ways of receiving love, providing specific examples of what that looks like to you.
4. Have you been receiving love from the things your spouse does for you that fall outside of your love language, or have you been missing out on noticing them?
5. Ask your spouse to describe one thing you could routinely do that they would appreciate and would help them feel loved.

Dear Lord,

Thank you for being a God of love and for making our love for each other possible. Thank you for not limiting our love and for showing us in every way how much you love us. Help us to give and receive love like you do.

In Jesus' Name, Amen

Are You Sure about That?

*"But above all, my brothers, do not swear, either
by heaven or by earth or by any other oath, but let
your 'yes' be yes and your 'no' be no, so that you may
not fall under condemnation"* (JAMES 5:12).

THE BOOK OF JAMES is one of the most practical books for building and maintaining relationships. In James 5, we find instruction for avoiding the Enemy's traps of dishonesty. His direction is simple: "Let your 'yes' be yes and your 'no' be no." In marriage, this means you need to be honest with yourselves and with one another about your commitments.

As a former people-pleaser, Chris might remind you of people you know or even yourself. He's hands-down one of the nicest guys you'll ever meet. However, early on in our marriage, he focused on saying what I wanted to hear instead of speaking honestly—he said yes to more than he could really commit to. His intentions seemed good; he was just trying to make me happy. But he'll admit he enjoyed occupying the pedestal I placed him on for doing what I wanted. It didn't take long for this to become a problem, as Chris often said "yes" when he truthfully meant "no." As life grew busier, he couldn't follow through on everything he committed to. I wasn't sure if I could take him at his word, and he began to lose his place on that pedestal.

Here's the reality. Whenever Chris didn't do what he said he would, it eroded trust and safety in our relationship. Without being dependable in the small things, I worried about his ability to come through on the bigger things. When I shared my concerns, Chris got upset and made excuses, but justifying his lack of follow-through just made things worse. Though he set out to make me happy, his people-pleasing nature backfired, causing him to end up doing just the opposite.

The solution James gives for this is straightforward but not easy. It might feel better or well-intended to say "yes" to your spouse now, then try to figure it out later. But as we learned, it's not a good long-term strategy. In fact, it's a dishonest strategy. Chris has worked hard to make his "yes" and his "no" meaningful and trustworthy. Even if I do get upset when he says he can't give me what I want—sorry, honey—he would always rather have my trust.

1. How do you struggle with saying "no" to each other? When do you risk your spouse being upset or disappointed with you?
2. When your spouse says "yes" to you, what makes you feel like you can trust what they say?
3. How do you react when your spouse tells you "no"? How can you improve in this area?
4. How would it affect your marriage if you said "yes" to each other all the time? What does God's Word say about this?
5. Can you identify an area in your marriage where you need to work on making your "yes" mean yes and your "no" mean no? Bring any areas to God and prayerfully ask him to reveal them to you.

Dear Lord,

Thank you for always being direct and honest. Thank you for also giving us the example and the encouragement to speak hard truths when we know that we are required to and that it is for the best. We ask that you give us a love that can withstand the difficult "no" answers. Help us to maintain honesty in our marriage and to never allow the Enemy to have a foothold.

In Jesus' Name, Amen

Let's Talk about It

"Know this, my beloved brothers: let every person be quick to hear, slow to speak, slow to anger; for the anger of man does not produce the righteousness of God" (JAMES 1:19-20).

GOOD COMMUNICATION is such a foundational part of a strong marriage. It always involves listening with the goal of understanding. Without these elements, you're almost guaranteed to find yourself in relational turmoil. How you communicate as a couple determines how connected you are, how well you can handle conflict, and how safe you feel in your marriage.

There are a few common mistakes married couples make when it comes to communication. One couple we worked with, Jacob and Jordan, kept making most of them. Whenever Jordan would bring up a tough topic to Jacob, she did so in an accusatory manner. To solve a problem, she wanted Jacob to know everything he was doing wrong, and her accusations would instantly shut Jacob down. Desperate to be heard, Jordan got louder and louder with her complaints, until Jacob stopped listening altogether.

On the other hand, Jacob attempted to solve problems by not addressing them at all. He withheld from openly communicating, assuming he knew how Jordan would respond. The longer time went on, the harder it became to keep his feelings stuffed away—so, when he finally expressed them, he'd erupt with anger. Between the two of them, healthy communication just wasn't happening. Jordan proved quick to speak, Jacob proved quick to anger, and neither one of them was listening to the other. This is the exact opposite of what God calls them to.

Remember, in marriage, you are on the same team. You are two who became one. Your spouse's issues should be important to you, and you should be willing to hear them out. When your spouse brings up a concern

Keys to Healthy Communication

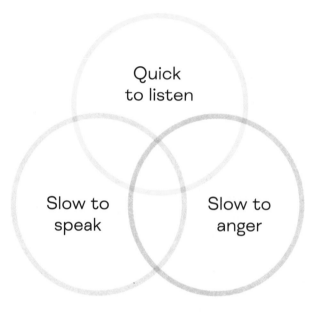

Quick
to listen

Slow to
speak

Slow to
anger

or frustration, lean in to listen instead of reacting. The conflict will only grow stronger and louder when your spouse feels unheard. Listening to understand is the best way to guard against anger, accusations, and repetitive messages. Let your spouse know that their struggle is valid and that you want to help find a solution.

1. How would you rate your communication skills as a couple, on a scale of 1 (poor) to 10 (healthy)? How would you like your spouse to improve their communication?
2. How would you describe the way you personally communicate? Why do you use that method? Does it work?
3. What did communication and addressing conflict look like between your parents in your home growing up? As a child, how did you

respond when your parents fought? Did you want to help fix the problem or hide in your room? Would you try to be good in order to shoulder some of the blame or keep the peace? How do these same behaviors or a rejection of those behaviors show up now as an adult?

4. When there is a conflict, how would you like your spouse to address it? What can you say or do to make communication go smoothly or to defuse the situation?

5. Examine the graphic, and discuss James 1:19-20 and Colossians 3:12-13. Which element do you need to improve on the most? Why do you think God tells us to do these three things? What happens when you do the opposite? What attitude should we bring with us into conflict?

Dear Lord,

Thank you for being a clear communicator with us. You are eager to hear us and slow to anger. Thank you for giving us all the grace we need in our marriage to learn the process of communication. Help us to treat one another with kindness, and help us learn to lean in and seek to understand one another instead of defending ourselves first.

In Jesus' Name, Amen

Made for More

"You are the light of the world. A city set on a hill cannot be hidden.
Nor do people light a lamp and put it under a basket, but on a stand,
and it gives light to all in the house. In the same way, let your light
shine before others, so that they may see your good works and
give glory to your Father who is in heaven" (MATTHEW 5:14-16).

AS A YOUNG COUPLE and new believers, we lived a comfortable life—predictable, safe, and filled with the busy monotony of our everyday tasks and duties. Neither one of us put much thought into the gifts we were given, much less use them to glorify God. But, as time went on, we matured in our marriage and our faith. We knew we were made for more. Individually and together, we could actually make an impact in this world.

Soon, we started having conversations about bigger, more important things—kingdom things. We began to ask God how he could use us for his purpose. By viewing our individual gifts, passions, and our story as something bigger than ourselves, we began the journey of helping married couples find redemption and fulfillment in Christ.

Perhaps it's time for you to do the same. Your marriage was made to shine before others and illuminate the glory of God in a dark world. This means giving up the predictable, comfortable lives you may have originally set out for. It means swapping baskets for stands, giving light to all in the house. Instead of choosing to exist horizontally within a self-focused life, venture into a vertical landscape, looking upwards and climbing uphill battles. You were made for more, and Jesus is worth striving towards a greater goal. There's a mountain out there, made for you to climb together, and it's up to you to pray and ask God to show you how. Kristen Welch, Christian author and founder of *Mercy House* and *Fair Trade Friday,* offers encouragement for climbing out of our comfortable lives: "Instead of telling God about the big mountain in front of us, let's tell the mountain about our big God."

If you are looking for a place to start, your light-shining, mountain-climbing moments may arise from the testimonies God has given you, combined with the passions he's placed inside you. He has allowed everything that is in your life purposefully. What God has helped you to achieve—and what you have yet to achieve—in your life and marriage is not meant to be hidden but, rather, to show the world his glory. What impossible thing can you accomplish together with God? The world is ready for his message to be displayed in the achievements made possible through your marriage. Decide to start shining your light together and see what amazing things he does through you.

1. Do you know what spiritual gifts God has given you? If not, look up a spiritual gifts assessment online, and take it. Share with one another your gifts and brainstorm ways you could use them.
2. How can your gifts complement one another?
3. What social issues break your heart? What problems upset you the most that don't seem to be getting better? What would it look like if this cause was a mountain you were made to climb?
4. Together, do you share a burden for a specific group of people? Who do you have an affinity for that you would like to serve or help?
5. Think of something you might be able to do as a couple to make a difference for the kingdom. It doesn't have to be big—just a start! Be willing to pray and ask God what he wants to use you for.

> *Dear Lord,*
>
> *Thank you for giving us a light to shine for your glory. Thank you for giving our marriage a purpose bigger than ourselves that will reveal your goodness. Help us to put your hope and glory on display for all the world to see.*
>
> *In Jesus' Name, Amen*

The Power of a Mustard Seed

"He said to them, 'Because of your little faith. For truly, I say to you, if you have faith like a grain of mustard seed, you will say to this mountain, "Move from here to there," and it will move, and nothing will be impossible for you," (MATTHEW 17:20).

AT A GLANCE, having faith the size of a mustard seed seems simple. How hard can it be to have just a little bit of faith, right? But, sometimes, gathering up even the tiniest amount is incredibly hard. It's all the faith we need to move mountains, but if we're honest, not many of us live like mountain movers.

With a lack of faith, Mariah and Ben found themselves feeling hopeless in their marriage. Right when Ben lost his job, his car happened to break down—and they were already behind in paying the bills. Meanwhile, Mariah struggled to support her mom, who was once again battling cancer. They started to view their problems as too burdensome for one another and too big for God to handle. In spite of their limited faith, Mariah and Ben saw God show up and move mountains in their marriage—Ben found a job and repaired the car, and though not fully healed, Mariah's mom was making strides in her recovery. Their hope wasn't restored because their circumstances changed; it was restored because they learned to depend on God together, no matter the circumstance. Faith does not require great outcomes—it requires trusting that God is who he says he is, and no matter what, he is enough.

In your marriage, there may be times when everything seems impossible, even mustering up a little faith. Remember that you serve the God of the impossible, and you truly can trust him with impossible things. When doors shut in your face, he can open new ones. When heartbreak nearly shatters you, he can repair and restore. When fears take your breath away, he can fill up your lungs with hope. Over and over again, he is a God who shows up and a God you can trust. In your weakest moments, he will act according to his faithfulness. So, like Mariah and Ben learned to do, choose to hang onto that little faith seed, knowing he's faithful. Together, take God at his word, and don't let go of trusting him to meet your needs.

1. When has it been the hardest for you to have faith that God would come through? Has he ever been unfaithful in your life?
2. What are you currently struggling to have faith about? What does the Bible say to do about this?
3. Look up Hebrews 11:1. How does this verse define *faith*?
4. Discuss a time when God did something that seemed miraculous, when you had no idea how a need would be met or a problem would be solved.
5. In general, what level of faith do you live by? Do you struggle to believe that God will come through for you, or do you never doubt him? Why or why not?

Dear Lord,

Thank you for being an on-time, ever-faithful, always-trustworthy God. Even in our doubt, you are faithful. We trust that you are capable of doing the miraculous with our little mustard seed of faith, which we try so hard to find sometimes. Help us to be a couple who trusts you and believes in God-sized goodness for our marriage.

In Jesus' Name, Amen

Wait for It

*"But if we hope for what we do not see, we wait
for it with patience"* (Romans 8:25).

IT'S SAFE TO SAY that patience is *not* my strong suit. Chris, however, *excels* in having patience. Truth be told, it's a little annoying, because his great patience highlights my lack of it—one of the many gifts of marriage. Still, his loving-kindness and patience towards me did two things: it gave me the space I needed to grow and start over, and it gave me an incredible working model of the power of patience.

His patience became something I longed to have, too. Patience is beautiful, especially when you're on the receiving side of it. Just imagine if God wasn't patient. A whole lot of us wouldn't have a chance to spend eternity with him; he'd quit waiting for us to repent and tire of forgiving us for the same sins over and over again. *Patience is beautiful.*

When it comes to your marriage, patience is a tremendous virtue, promoting forgiveness, kindness, gentleness, and encouragement. It says, "I will wait for you to grow, learn, fail, recover, and even to run back inside to get your phone you forgot when we're already late." Patience changes people, helping us to become more like Christ. Patience helps you wait for whatever God is doing in your spouse, giving them the room they need to hear his voice and respond to his words. Patience encourages you to wait on others with a genuine sense of serenity, and hope in all the good things to come.

When we look at some of the alternatives to patience—pressure, judgement, accusations, fit-throwing, and demanding—we can understand how damaging impatience can be. Not a single one of these will benefit your marriage. A lack of patience bears a crushing weight, creating an atmosphere of shame, chaos, and frustration. There's a reason why patience is so hard for so many of us—it's packed with life-changing power. Despite the warnings that God might trick you into waiting, you actually might want to pray for patience.

As for the two of us, Chris is still the more patient one, but in God's grace, I've come a long way—and so can you. A spouse's patience allows for God's plan to unfold in your marriage. Put patience on display together. All the good things truly are worth waiting for.

1. Which one of you is more patient? Be honest.
2. Is having patience something you struggle with? What are some ways you can begin growing more patient?
3. In what ways could a lack of patience negatively affect your marriage?
4. Do you show more or less patience to your spouse than to other people (i.e., other drivers on the road, people waiting in line, coworkers, friends, and so forth)?
5. What would your life be like without God's patience? How do you think he models patience for us?

Dear Lord,

Thank you for giving each of us the patience we need. Thank you for waiting for our repentance and meeting us in our sin, not with shame, but with tolerance and forgiveness. Help us to wait with hope for all that you have for us, as we grow more and more like you.

In Jesus' Name, Amen

Let's Connect

". . . and may the Lord make you increase and abound in love for one another and for all, as we do for you . . ." (1 Thessalonians 3:12).

PAUL WAS EXCITED and comforted when he found out from Timothy that the Thessalonians had continued to live in faith and love. He expressed his earnest longing to return to see them in a report of his prayers. Paul desired to reconnect face-to-face with the Thessalonians, hoping to supply what they needed in order for their faith to grow (1 Thessalonians 3:10). Paul's connection with the church strengthened their unity and provided for their needs. And this is exactly the role of connection in your marriage.

Without effective communication, connection is impossible. In your marriage, communication is essential in bringing you closer together. When one couple we worked with, Josh and Caroline, hit a rough patch, Caroline began to withhold her thoughts and feelings from Josh. She assumed her emotions would be obvious to him, so when they weren't, she began to pull away from Josh in frustration. Her lack of communication kept Josh from connecting with her. As time went on, Caroline began to build resentment—the only thing more dangerous to connection than a lack of communication.

In order for Caroline and Josh to reconnect, Caroline needed to learn how to share her thoughts and feelings with him. Once she could be honest about her emotions, she could begin to let go of her resentment. And, as Caroline opened up, Josh could bring what she needed to their relationship, just like Paul sought to do for the church in Thessalonica.

As a couple, if you aren't accessible or engaged in honest communication, there won't be opportunities for connection. Don't assume your spouse knows what you're thinking or how you're feeling—communicate it. As you learn to connect, you'll also notice when disconnect occurs. That's because connection is so fulfilling, you'll long to regain and sustain it. If you want to stay connected, prioritize communication. Know that when you communicate, you connect, and when you connect, you increase and abound in love for one another.

1. Have you ever had someone stop talking to you? If so, how connected did you feel to that person?
2. What situations, emotions, or stressors in your marriage cause you to want to disconnect from one another?
3. When are the times you feel most connected to one another? Describe what that feels like. How do you know if you are disconnected? What does that feel like?
4. What do you think would happen to the couple who gradually becomes disconnected over time? What would that slow fade look like?
5. How can you be sure to stay connected? Discuss practical steps you can take when you sense disconnection is taking place, whether you are the one pulling away or your spouse.

Dear Lord,

Thank you for being a God who cares so deeply to connect us to yourself and to each other. Thank you for filling up the Bible with so many "one another" commandments. Your heart for your children to remain together is so evident, and we believe you desire for us to be especially connected as husband and wife. Help us to remain faithful in communicating, so we can remain powerfully connected, and thereby excel in caring for each other's needs.

In Jesus' Name, Amen

To Love Is to Be Vulnerable

"But he said to me, 'My grace is sufficient for you, for my power is made perfect in weakness.' Therefore, I will boast all the more gladly of my weaknesses, so that the power of Christ may rest upon me" (2 CORINTHIANS 12:9).

SHAYLA WAS a child of divorce. She grew up in a hostile home environment with her mother and her stepfather, who constantly fought and rarely attended to her needs. Shayla learned from an early age that she had to depend on herself. She was forced to be strong and to never show any weakness, because no one ever helped her when she did.

One day, however, Shayla met Ben, and his abundant kindness swept her off her feet. He attended to her needs, and she found herself feeling safe in his presence—so safe that she said *yes* when he proposed. Her comfort with Ben grew throughout their dating and engaged relationship, but things changed when they got married. After moving in together, Ben really began to see Shayla's flaws. This made her feel vulnerable, continually afraid that Ben would leave her at any moment. She started to get angry easily, grew increasingly defensive, and looked for fights. It seemed like she was trying to get him to leave.

Why would Shayla try to push away a man she loved so much? Because his love for her made her feel vulnerable—to rejection, to being found out for her flaws, to being abandoned and uncared for once again.

You cannot have real love without vulnerability, nor love without grief. Feeling one precipitates feeling the other. And this scared Shayla. Fortunately, Ben knew Shayla needed help facing this fear. They sought counseling and began to process Shayla's painful past. She learned to open up to Ben, which allowed him to see the pain and insecurities she had perceived as weaknesses that needed to be hidden and guarded. Through all of this, Shayla learned that sharing her weaknesses and struggles with Ben allowed God's grace to pour into their relationship. The safety she found in opening up and exposing her vulnerability ultimately created an unbreakable closeness in their marriage.

1. What are some weaknesses you think you have? Ask your spouse what they think about them.
2. Do you see weaknesses as something you need to perfect? Do you think it's necessary to work on all of your weaknesses, or are there some you're okay with? What does God say about weaknesses?
3. What makes you feel the most vulnerable? Are you willing to trust your spouse with this vulnerability?
4. Sit across from one another—knee to knee and eye to eye. Set a timer for four minutes. Stare into each other's eyes for the entire time without talking. (Really—be willing to move through the giggles and the awkwardness.)
5. Discuss what that exercise was like for you. How did it make you feel? Why do you think it made you feel that way?

Dear Lord,

Thank you for providing sufficient grace for our shortcomings and for using them as opportunities to display your glory. Help us to be vulnerable and never hide our weaknesses from one another. Instead of being ashamed of our weaknesses, let us be a couple who sees and knows your strength in them.

In Jesus' Name, Amen

What Are You Waiting For?

"And now, O Lord, for what do I wait?
My hope is in you" (PSALM 39:7).

WHEN YOU ENTERED into your marriage, you brought some things with you, including all your "if/then" beliefs. That is, "if I do things for my spouse, then my spouse will always show their appreciation for my efforts," which may or may not be true. In other words, you brought expectations. We all have them, but we don't always realize it until they go unmet and wreak havoc.

Expectations are our ideals—how we believe things should be. Some marital expectations are healthy. For instance, you should expect your spouse to remain faithful, to speak kindly, and to not abuse you physically, mentally, emotionally, or spiritually. However, as a married couple, you'll learn how to leave many of your damaging expectations behind. When we expect things from our spouse they cannot provide, especially things that only God can provide, our expectations are truly just premeditated resentment.

Only God provides salvation, rescue, peace, joy, perfection, and fulfillment. He will one day heal every pain and restore every brokenness. Your spouse may help alleviate the burdens of pain and brokenness in the meantime, but you can only expect Jesus to be your hope. Expect your spouse to be God, and you will find resentment. Expect God to be God, and you will find fulfillment. Marriage gets better when we set realistic expectations for one another, which requires placing our ultimate hope in the Lord.

After a friend deeply hurt her feelings, Kimber expected Sean to know exactly how to comfort her. But after bringing up her pain, he did not respond according to her expectation. Instead of affirming her and offering an embrace, he told her that she might be overreacting and to consider her friend's point of view. Disappointed, Kimber thought to herself, "If he cared about me, he would understand how I was feeling and try to help me feel better." Notice the "if/then" thinking—these are danger words. When those two little words enter into your marriage vocabulary, it's time to assess your expectations. Are they realistic for your spouse? Perhaps instead of placing expectations, notice your desire within it, and share what you need openly with your spouse.

When you quit longing for your spouse to deliver something they cannot give you, you will give your marriage one of the best opportunities for happiness.

1. Besides the expectations we mentioned, what are some healthy expectations you can have in your marriage? What are some unhealthy expectations, or premeditated resentment, you can create in your marriage?
2. Think of a time when your spouse has let you down. Do you recognize an unmet expectation you had in that situation? Did you notice any "if/then" thinking?
3. When your spouse doesn't meet your expectations, how do you typically respond? Is there an opportunity for growth here? How can you respond instead?
4. Did you have expectations placed on you as a child? What were they, and how did you feel about them?
5. How do you think having unrealistic expectations on your spouse can cause them harm? How can it harm you?

Dear Lord,

Thank you for being a God who does not disappoint or fail us. You are the only one capable of meeting all of our needs. Help us to only have realistic and healthy expectations of one another and, in all things, to place our expectant hope in you.

In Jesus' Name, Amen

Putting Anger to Bed

*"Be angry and do not sin; do not let the sun go
down on your anger, and give no opportunity
to the devil"* (EPHESIANS 4:26-27).

WHEN YOU GOT MARRIED, someone probably offered you the advice,
"Don't go to bed angry." But is that really a good practice? Well, kind of. In
Ephesians, Paul is not telling us to resolve every fight and move past all our
anger before bedtime. But he is saying that you shouldn't dwell in your anger
or leave it undealt with. Anger deserves addressing.

Let's be clear—you *are* allowed to be angry. Anger itself is not a sin, but
how we deal with it can be. And not dealing with anger at all is a problem, too.
Avoiding your anger gives it an open invitation to linger, which invites the
Enemy to have his way with it. In your marriage, undealt-with anger causes
bitterness and resentment to take root, casting darkness in your relationship.

Recurring anger in your marriage requires attention. Anger always has a
message, but unleashed inappropriately, the real message gets masked in the
explosion. In general, anger is a secondary emotion—it comes with another
emotion underneath it. Quite often, those other emotions involve hurt, fear,
frustration, inadequacy, or guilt; they may be much more difficult to feel,
which makes them less accessible than anger. After all, anger makes us feel
powerful, but hurt and inadequacy make us feel vulnerable.

Becca and Jared had just mortgaged a house and bought a new car. When
they did, Jared had a reliable, well-paying job. Lately, however, there had been
several layoffs at his office, and he started coming home in a bad mood. He
would snap at Becca and even yell at their dog. Becca wondered why he didn't
seem happy anymore and why they were fighting more.

One night, after a big blow-up between them, Jared broke down in tears.

The root of his anger pushed itself through the surface—he was afraid. He was scared of losing his job and the ability to provide for his family. Feeling anger seemed easier for Jared, but doing so it only covered up his deeper fear. Once Jared opened up to Becca, his anger began to subside.

As a couple, whenever anger rears its head, look for a deeper emotion it might be covering up. Sometimes, you just need to control how you deal with your anger, but other times, you need to make sure to hold your anger up to the light, so that the real problem can be seen and addressed together.

1. When do you typically begin feeling angry? How quickly do you tend to move on from anger? Can you identify the underlying emotion triggering your anger response?
2. When is the last time you experienced anger turning into resentment? How could holding onto anger in your marriage affect the way you feel about your spouse? What does God's Word say about it?
3. As a child, were you allowed to show anger or were you made to bottle it up? How were you taught to express your frustration, disappointment, fears, and so on?
4. Is there any undealt-with anger in your marriage right now? If so, what is the issue, and what can you do to resolve it? How can you sow peace back into your marriage? (See Romans 12:18, 14:19.)
5. Come up with a game plan for how you will deal with anger, whether you are the one feeling angry or whether it be your spouse. What is an appropriate response and what is not, in your opinion?

Dear Lord,

Thank you for being a God whose anger is only righteous. Thank you for allowing us to experience all emotions, including anger, and thank you also for giving us the gift of self-control. Help us to use that gift abundantly in our marriage.

In Jesus' Name, Amen

You Can Do It!

"Therefore encourage one another and build one another up, just as you are doing" (1 Thessalonians 5:11).

MORE THAN ONCE in our office, we've heard a wife badmouth her husband. She'll refer to him as an extra child or describe him as incompetent with specific tasks, and it instantly creates an awkward atmosphere. The same applies when a husband openly complains about his wife, or sarcastically refers to her as "the boss" with whom he needs to clear every move he makes. Neither of these are acceptable ways to speak to one another or about one another in marriage.

God calls us to a higher standard. He calls us not just to avoid speaking poorly of one another, but to speak with encouragement and affirmation. He asks us to speak with the purpose of building up, not ridiculing or disregarding, especially when in public.

Encouragement can bring great gifts for your spouse. Your kind and generous declarations can build their confidence, helping them to boldly step out in faith, even when they feel afraid or inadequate. Your words may serve as the exact assurance they need to dream bigger, live abundantly, and chase down their calling. At the very least, your encouragement can champion them, as they continue seeking God's goodness.

While speaking affirmation lets your spouse know you believe in and trust them, it also gives them permission to fail. When I moved our family across state lines to a town where they had no friends, no family, and no job, Jamie's simple words, "You're doing the right thing," allowed me to take this bold step. Sometimes, all your spouse needs is to know you're on their team—that they aren't risking your support when they take risks for the gospel.

Without Jamie's validation, I may have stepped down instead of stepping out in faith, and we would have missed out on the lives we have now.

When you encourage your spouse, you build them up. When you fail to, or worse, when you speak negatively about or to them, they'll feel easily torn down. Be your spouse's biggest fan. Everyone needs someone to believe in them, and when it comes to your partner, you are the best person for the job.

1. How does it make you feel when your spouse tells you they believe in you?
2. Is there an area in your life that you're feeling discouraged in right now? How do you typically share your discouragement with one another?
3. When God tells us to encourage one another, do you think this means all the time? What does encouragement look like to you in both challenging and fulfilling times?
4. To what extent does your personality thrive on encouragement? How much affirmation do you tend to need when making decisions or taking risks?
5. Spend time sharing with one another five things you think your spouse is really good at. How can you encourage them today?

Dear Lord,

Thank you for giving us each other as a support system for when we feel insecure or discouraged. Thank you for desiring us to be built up, so we can each become who you made us to be. Help us to avoid any discouraging or condescending talk, especially in front of our friends and family. Teach us to be our spouse's biggest fan.

In Jesus' Name, Amen

I Get It Now

*"Rejoice with those who rejoice, weep with
those who weep"* (Romans 12:15).

ONE OF THE GREATEST TOOLS for any relationship is empathy. Putting
yourself in your spouse's shoes allows them to feel seen, heard, and understood. Showing empathy invites your spouse to take a deep breath—a sigh of
relief—in knowing they are not alone.

Eric and Marissa ended up in quite a few fights, and as time went on,
the fighting became more frequent and intense. Marissa was raised by an
alcoholic father, and life was hard. However, the difficulties she experienced
happened in the past, and Eric didn't connect it with the present situation.
All he knew was every time they got into a quarrel, Marissa grew strongly
defensive or left the room altogether. Eric had no idea why she couldn't have
a heated discussion without things going south very quickly, and his frustration with Marissa amplified.

After one major argument, Marissa finally revealed that whenever Eric
raises his voice, it reminds her of her father, who would come home drunk,
raise his voice, and lose his temper in terrifying ways. The sound of an
elevated voice ignited fear in Marissa, leaving her ready to fight or flee. Eric
sought to better comprehend her feelings and experiences and make sense of
her behaviors in their marriage. He now understood that Marissa wasn't just
responding to him; she was reacting to fear. Without mutual understanding
and empathy, they would have never been able to move past these episodes.

Practicing empathy allows us to connect with each other. Eric could have
explained to Marissa that he wasn't her father or treated her reactions as
dramatic or silly, but that wasn't what she needed, which was to be understood and free to own her feelings. Now, Eric is able to meet her needs by
considering the volume of his voice, and reassure her that he was on her side.

A little compassion in your marriage can go a long way. The goal is not always to agree but to understand.

1. Can you identify any triggers in your marriage, like what Marissa experienced with Eric's raised voice?
2. We often want a quick solution so things can go back to "normal," but sometimes, the fix is just to listen and attempt to understand (1 Peter 3:7). What are your thoughts on this? Have you experienced this tension in your marriage?
3. How do you feel when your spouse doesn't understand what you're telling them? What can you do to help them understand? Is this something you can ask the help of the Holy Spirit for?
4. Think of a childhood memory that has an emotional impact (i.e., a pet dying, a divorce, bullying, and so on). Share your memory with your spouse, including any details you remember—how old you were, where it was, when it happened, and how it made you feel. Have the spouse who listened to the story retell it entirely, *as if it happened to them*. Repeat the exercise until you both have a turn sharing, listening, and retelling.
5. How did it feel to share your story? How did it feel to hear your spouse's story and repeat it back to them?

Dear Lord,

Thank you for allowing Jesus to come live amongst us and experience the things we experience. Thank you for being a God who understands our pain. Please give us the ability to see the pains we each carry and the ability to understand them. Help us to learn to weep and rejoice with each other, and to bring healing to the areas where pain has been residing.

In Jesus' Name, Amen

The Freedom of Forgiveness

"Put on then, as God's chosen ones, holy and beloved, compassionate hearts, kindness, humility, meekness, and patience, bearing with one another and, if one has a complaint against another, forgiving each other; as the Lord has forgiven you, so you must also forgive" (COLOSSIANS 3:12-13).

FORGIVENESS COMES UP a lot in our counseling office. It's often a topic that needs to be addressed among victims of trauma, adult children from abusive families, and married couples with all kinds of stories. And forgiveness is hard—it's nothing like any of us often hope it will be. Forgiveness rarely feels fair or freeing, at least at first. Naturally, our biggest struggle with forgiveness is not the forgiving itself, but accepting the arduous process of forgiveness.

We all want forgiveness to be earned and an instant release of the damage done to us. But forgiveness is challenging—even Jesus will tell you that, and no one knows better than him. We have done *nothing* to earn the forgiveness Jesus gave us through the cross, yet he gives it freely. The Father intended his Son to bring forgiveness, and he still intends forgiveness for us today. At its very core, forgiveness provides freedom. If you want true freedom in your marriage, you need hearts that forgive.

When Zach found out Brittany had been lying to him for months about her spending habits, he was furious. After working diligently to get them out of debt, finding out that she sabotaged his efforts behind his back was painful. They had a long, heart-to-heart talk, and Zach chose to forgive her, but he didn't forgive naively, for forgiveness does not automatically equate to trust.

Forgiveness is a process, and although Zach took the first step to forgive Brittany, he didn't immediately trust her again. The act of forgiving set him free from building resentment and focusing on the pain, but it didn't make her trustworthy. Her deception had caused damage, and they needed to develop a new history.

In your marriage, it's important to remember that forgiveness is a process. First, it involves a choice to forgive, then a daily decision to walk in forgiveness. Finally, forgiveness requires time to let feelings dissipate. When your spouse wrongs you, which they will, choose to show them grace, continue walking in your choice, and while your spouse changes their behavior and rebuilds your trust, allow feelings of forgiveness to freely flow. Forgiveness doesn't mean that what your spouse did was okay; it means you are free from the burden of not forgiving, and you can learn to move forward together.

1. Do you find it hard or easy to forgive someone, especially your spouse? Why or why not?
2. Do you have anything you need to ask your spouse for forgiveness for right now? If so, do it.
3. Read Matthew 6:14–15. What do you think this verse means?
4. What harm does it cause *to you* when you choose not to forgive your spouse or someone else who wrongs you?
5. Ask God to reveal to you if there is anyone in your life right now that you need to forgive. Decide what steps to take to work toward forgiveness. Are you ready to do that?

Dear Lord,

Thank you so much for pouring out your forgiveness on us and for dying for our sins, even while we were still sinning. Thank you for commanding us to forgive so that no barrier can divide us as husbands and wives. Help us to have hearts that are cautious against hurting one another and willing to be merciful when we do.

In Jesus' Name, Amen

Do You Smell Smoke?

*"So also the tongue is a small member, yet it
boasts of great things. How great a forest is set
ablaze by such a small fire!"* (JAMES 3:5).

AS A YOUNG GIRL playing in my mother's bathroom, I came across the lighter my mom used to melt the tip of her eyeliner, along with a recent picture I had colored for her. Out of curiosity, I took the lighter and held its small flame to the corner of the paper. Almost immediately, the entire page lit up. A nearby towel soon caught fire, which proceeded to set the wallpaper ablaze. This tiny flame took out an entire bathroom within seconds. Who knew such a small spark could cause such a big disaster? As a child, I had no idea. But God does.

God compares our tongues to this spark. Just as that lighter's flame spread within seconds to the whole bathroom, the words that spark off our tongues can cause great damage in a short timeframe. Can you think back to a time when you were called a name or insulted? Is there something some- one said about your personality or your appearance that you can still feel the sting from, even though it's been years? That's evidence of the power words have in our lives.

When God made all of creation, he *spoke* it into existence—he used his words. Your words have the power to speak life or destruction in one another. God calls us to speak life and to control our tongues. Remember when we are told to be "slow to speak" in James 1:19? This is why—uncontrolled words cause harm. And not all of the damage comes with fire insurance. The dam- age we cause others with our words is not always reparable like my mom's bathroom. Imagine your husband or wife carrying around pain from words you said years ago, just from one angry moment.

In your marriage, treat one another tenderly with your words. Let your voice be soft and thoughtful, and kinder than the hurtful ones your spouse may have heard throughout their life. Sticks and stones may break our bones, but, according to God, words definitely *do* hurt us.

As a believer and follower of Jesus, you are called to speak truth, hope, and encouragement, not destruction, to all of those around you. So how much more should you do this with the one you married?

1. Talk about a time when you were hurt by someone's words. What were they? Do they still cause you pain?
2. Read Proverbs 12:18 and 21:23. Do you tend to speak before thinking? If so, what makes this hard for you? What are some ways to hold your words back?
3. How would withholding harsh words change the outcome of an argument you might have with one another? Why do our harsh words hurt so much?
4. Have you ever said hurtful things to each other? What were they? How did it affect you?
5. What do you think God means by life-giving words? How can you begin to use your words for good in your marriage? How are you doing at accomplishing this every day? Are you willing to make life-giving words a part of your daily life?

Dear Lord,

Thank you for this powerful visual of setting forest fires with our words. We trust that this truth is meant for our good and for the purpose of caring for one another as speakers of life and hope, not of destruction. Help us to be a couple who pauses before we speak, so that we may use our words for one another's good.

In Jesus' Name, Amen

Whose Fault Is It, Really?

"So then each of us will give an account of himself to God" (ROMANS 14:12).

ARE YOU FAMILIAR with the game you can never win? Believe it or not, it's a pretty popular one. In fact, couples all over are playing it. It's called the blame game. Here's how it works—accuse one another of anything you want without taking any personal responsibility, and there you have it—everyone's a loser. Sounds fun, right?

We've seen the blame game played out a lot, and Isabelle and Tyler were professionals. They had been practicing for years, losing a healthy marriage in the process. Tyler consistently ignored any of Isabelle's requests for help, and Isabelle consistently blew up with anger. However, Tyler insisted that if Isabelle would just ask him to help out without nagging, he wouldn't *have to* ignore her. Likewise, Isabelle maintained that she wouldn't *have to* keep blowing up if Tyler just listened to her when she asked for help.

Isabelle and Tyler didn't understand that your spouse should not be held accountable for your behaviors—only you should. God does not want us to blame others for our responses. In your marriage, this means no matter how your spouse acts, you are in control of how you respond. Isabelle was responsible for communicating without belittling her husband no matter how much he ignored her. And it was Tyler's duty to offer help and stay engaged when his wife spoke, no matter how much she hassled him.

By taking personal responsibility for your actions, you begin to take control of them and relinquish a sense of controlling your spouse. Practice owning your responses, and leave the rest up to your spouse and the Holy Spirit. The Holy Spirit can then do his part to convict Isabelle of the tone she used. If Isabelle spoke patiently, even as Tyler ignored her, the Holy Spirit would convict him as well. God does not honor excuses for avoiding ownership of your actions. Your spouse should not control your actions, and you should not control theirs, either. We all need to take personal responsibility for our own actions.

1. Do you tend to blame your spouse for your negative behaviors? Has blame ever been a factor in your marriage? How has that worked out?
2. God doesn't call us to control one another, but we do have *influence* over each other. What could Isabelle and Tyler have done to potentially influence their spouse to respond more adequately?
3. Do you see any areas in your marriage where you could influence your spouse more positively?
4. How could taking personal responsibility for the things you do and say invite the work of the Holy Spirit into your marriage? How have you witnessed him working in you or your spouse?
5. Is taking personal responsibility difficult for you? Why or why not?

Dear Lord,

Thank you for the gift of self-control and the ability to take ownership of our responses. Even though this can be painful, it also allows us to change. Help us to own what is ours to own and to leave the Holy Spirit to do the rest. And when he does, help us to praise you for it.

In Jesus' Name, Amen

Everything I Do, I Do for You

*"Do nothing from selfish ambition or conceit, but in
humility count others more significant than yourselves.
Let each of you look not only to his own interests, but
also to the interests of others"* (PHILIPPIANS 2:3-4).

NOT MANY RELATIONSHIPS will call you to selflessness like marriage will. And not many relationships will bring out your selfishness as much as your marriage does. What a winning combination—a call to put your spouse first amidst a strong desire to be selfish! Marriage certainly offers plenty of opportunities to respond to the needs of our spouse, especially when we may have our own needs that we would like met first. It is in those times of sacrifice that we are being given the opportunity to become more like Christ. It's a chance to die to ourselves (Galatians 5:24).

During his time on earth, everything Jesus did was with the purpose of glorifying his Father and restoring our relationship with him. Of all the ways God could have demonstrated his love for us, he chose sacrifice. Jesus could have easily come down from the cross and saved himself, but he didn't. Instead, he deliberately made the greatest sacrifice of all for us as the Son of God. He didn't have to leave the glory of heaven, be born in a manger, live humbly as a carpenter, or take even one step on this earth with us. It's overwhelming to understand that instead of avoiding suffering or sacrifice, Jesus came towards it willingly because he loves us.

This is the kind of love Jesus calls us to in marriage—a selfless love that puts your spouse before yourself and has you thinking of your spouse's needs before you consider your own. Your marriage will work best when the two of you are willing to sacrifice for each other. When you walk in the way God calls you to, you can be assured he will supply everything you need. As you learn to love one another sacrificially, looking out for each other's best interests, you can trust you'll become more like Christ each and every day.

1. Do you feel like your needs are equally met in your marriage? Is this an area you need to work on?
2. Does Philippians 2:3-4 mean you should only focus on your spouse's needs and not have a voice in your marriage for your own? Why or why not?
3. How do you feel about asking your spouse to meet a need that you have? What are some ways you can address this? Do you feel hopeful about your spouse seeing your perspective on this and coming through for you?
4. Are you able to negotiate or meet in the middle? How could you add more compromise to your marriage so that all of your needs could be met?
5. Share a current need that you have, and tell your spouse what they could do to help meet that need.

Dear Lord,

Thank you for the gift of your Son, for not only sending Jesus, but for planning for him to come and pour himself out in agony on a cross for us from the beginning. Thank you, Jesus, for the amount of suffering you went through to take on our sin and shame. Help us to not take that gift for granted, and to demonstrate the same kind of selfless and sacrificial love in our marriage.

In Jesus' Name, Amen

What's the Difference?

*"Now you are the body of Christ and individually
members of it"* (1 CORINTHIANS 12:27).

HOW LONG were you married before you realized your spouse was weird?
They load the dishwasher all wrong, put the toilet paper roll on upside-down,
and get bothered by stuff you find trivial or ridiculous. All of these things that
make your spouse lovable or obnoxiously weird, depending on the day, also
make your spouse different from everyone else. They're part of what makes
them exactly who God designed them to be.

In the church of Corinth, everyone had a role. There were teachers,
administrators, encouragers, and givers. And no role was more important
than another. Each member served a necessary purpose to help the church
function at its best. This is still how the church works today. Similarly,
society functions well when everyone is valued for their differences, free to
operate within their design, and valued as a person altogether. Even more,
this is also how your marriage works best. God knew exactly what he was
doing when he made us different.

Our differences aren't limited to just our spiritual gifts. People vary in
assertiveness, in extroversion, in appearance, in what makes them laugh
or cry. We're all different letters, colors, animals, or numbers—whatever
personality test you like best. We are different in so many ways, and each one
of us is unique. God is wildly creative, but his design for diversity is not just
inventive for the sake of being inventive—it's purposeful.

God gifted people with a propensity for varying strengths and tasks
within the Church for a reason. That way, as we each fulfill a different role,
the Church can expand and grow. Desire this kind of growth in your marriage. The ways your spouse differs from you can help you grow where you

need refinement, learn where you lack perspective, and fulfill a role you cannot fill by yourself. Your spouse's differences are a gift not a liability, and not something you should seek to change. Embrace each other's uniqueness, and take your marriage to an entirely new level you'd never reach without a little weird.

1. What does your spouse do that you think is odd or even bizarre? Have a little fun sharing with one another.
2. Are there areas where you wish your spouse was more like you? Why might your differences be good? What purpose does God have for your differences?
3. How has God hardwired you? Name some specific traits your spouse brings to the marriage you don't have that makes your marriage work better. What would your marriage be like without those specific traits?
4. Name three ways you're similar and three ways you're different. Do you see these things as good, bad, or both?
5. For fun, take an Enneagram assessment and see what you learn about yourself and your spouse. Compare results together, and start a conversation about what you learned.

Dear Lord,

Thank you, God, for being so creative and purposeful. There's not one of us made by mistake or with error. Thank you for using every quirk and difference we have to help make our marriage stronger, allowing us to become more like Jesus as we grow uniquely together.

In Jesus' Name, Amen

Keep Going

"And let us not grow weary of doing good, for in due season
we will reap, if we do not give up" (GALATIANS 6:9).

COMMITMENT IS AT THE CORE of every accomplishment. Without dedication
to a cause, there's not much you will achieve or complete, including your
marriage. If you want to make it until death do you part, you need to stay
committed. Commitment isn't easy, but God encourages us graciously to
push through the weariness, reminding us that a time of reaping the reward
will come.

For years, Kendra waited for Brandon to connect with her. When they
were dating, they talked constantly and told each other everything, but they
drifted apart after getting married. Brandon seemed more stressed, and
he started to change. He spent way more time with his friends than with
Kendra. Whenever Kendra did have time to talk with him, his answers were
short, and he just seemed disinterested. Kendra grew weary in their relationship and began to think that things would never change.

With the wise counsel of a friend, Kendra decided to stay the course. She
spent more time praying and less time worrying and striving, continually
asking God to soften Brandon's heart towards her. Every time he chose to go
out with friends, she prayed for him to return. After a solid year of committed prayer, Brandon came home early one night. When Kendra asked him
if anything was wrong, he told her that he just didn't feel like hanging out
with the guys anymore. This was the first glimmer of hope she had seen in a
long time.

Several nights passed by and Brandon continued to stay home. As those
nights progressed into weeks and months, Kendra and Brandon began
spending more and more time together. They started watching movies,

cooking fun meals, and even training to run a marathon together. After all those years of staying committed, the season of reaping finally came.

Your marriage will not always be easy, but it also won't always be hard, either. When those hard stretches come—which they almost inevitably will—and weariness sets in, stay committed to your vows, invite God in, and remember reaping will come in the end.

1. How committed are you when things get hard? Do you have a tendency to give up early?
2. Do you find it difficult to continue doing good while you're waiting on God? Do you ever run ahead and try to fix things on your own instead?
3. Do you have a plan in place for how you will stay committed as a couple?
4. Has anyone in your life ever given up on you before? How did this affect you? Do you ever feel like God has given up on you?
5. What do you think would have happened if Kendra gave up on Brandon? What are some things that might cause a couple to drift apart like Brandon and Kendra did?

Dear Lord,

Thank you for being a God who rewards us at the end of our commitment. A relationship with you is our greatest reward. Thank you for not ever giving up on us and for constantly pursuing us. Help us to never give up on one another, take our vows seriously, and remain completely committed.

In Jesus' Name, Amen

Come on Over

*"Contribute to the needs of the saints and seek
to show hospitality"* (ROMANS 12:13).

SOME DEAR FRIENDS of ours, Jon and Sharon, have always kept an open-door policy in their home. They've never claimed their home as their own, and through the years, we have watched them share it with countless people. Whether someone needed a place to crash, throw a celebration, or call home while they served in ministry, their doors were open. They intentionally made their home the biggest vessel of their ministry together.

God calls us all to the ministry Jon and Sharon model so well—meeting the needs of fellow believers and showing hospitality to the lost. At the end of each of the many soccer seasons that Jon coached, he would host the whole team at their house for pizza, celebrating the individual gifts of each team member by naming them in front of their peers. Sharon uses every opportunity she has to open her doors as well, whether it's hosting couples for a Bible study, offering neighbors a place to connect, or helping international friends trying to learn English. Jon and Sharon live a completely missional life, often without leaving their home. This is gospel, God-centered living.

You are called to be givers, taking what you have and using it for God's glory. Your home offers a perfect platform to do this from, but it certainly isn't the only way. You aren't Jon and Sharon, and the neighbors you invite over might not be soccer teams or international peers. But you are who you are and where you are for a purpose—to love God and to love your neighbors.

As a couple, seek to show hospitality in whatever ways you can. Remember that gospel hospitality is not confined to your home address. It bids us to open up our hearts, make a place at the table, and raise your voice to say, "You belong here." Use your marriage as an invitation for others to find a place in God's family, where they're always wanted and forever belong.

1. Have you ever thought of using your home as a place of ministry? What are your thoughts about this? Do you feel like this might be a gift you have to share?
2. Jon and Sharon used their home for Bible studies, sports parties, international neighbors, and missionaries. Who could you invite into your home? Where could you find people to invite?
3. What do you think gospel hospitality is? Give some examples from people you know or from Scripture.
4. What other ways can you support local believers and engage with people who don't know God? Who is God putting on your mind right now?
5. What has God given you that, together, you're willing to surrender to him? What can God use of yours for his glory?

Dear Lord,

Thank you for giving us all that we need to contribute to your ministry and to show hospitality. Thank you for providing gifts to each of us, so that we can give back to those in need. Help us to surrender all that we have to you, using what we have as you desire.

In Jesus' Name, Amen

Just Hang On

*"Now faith is the assurance of things hoped for, the
conviction of things not seen"* (HEBREWS 11:1).

WE'VE CERTAINLY HAD our fair share of hard times, where all we had to hold
onto was hope in God. I suffered many illnesses and underwent numerous
surgeries. Chris went through a big health scare and a few company layoffs.
We've experienced financial strains, pregnancy complications, a miscar-
riage, and more than our fair share of loss and grief.

These kinds of hardships hit unsuspecting couples everywhere. Not long
ago, we sat across from a couple on the brink of divorce after losing their son.
They each experienced and expressed grief in completely different ways. And
neither of them knew how to go through this devastating loss, having never
experienced anything close to this level of pain before. They could barely
survive it on their own, much less help their spouse through it.

As we walked this couple through grief, we gave them permission to experience their pain however they needed to. Hard times that include hurt, fear,
and anxiety necessitate an abundance of grace. When you experience a trial
as a couple, it's okay if you have no idea how to handle it. When your hearts
grow overwhelmed, your focus and clarity get foggy. That's a normal part
in the process of suffering. Knowing this, and allowing space to feel overwhelmed, is half the battle.

The other half is fighting to have faith in unseen goodness, comfort, and
the work of the Father. While your emotions go haywire, remember that his
steadfastness does not. Cling to him when your circumstances throw you
into the raging sea. He is the anchor for your soul, as the stormy waters crash
over you. Hold on long enough, and you will soon find your way again.

No matter what trials you face in your marriage, you will never face them alone. The fix may not come easily, but hope is always there. If you drift apart during the raw emotional stage of survival, that's okay—hang on, as the Lord calms the storm and brings you together again. Always be willing to grieve, riding out the pain while you await the relief he promises. Be authentic, talk about how you feel, and cling to your faith. The unseen is waiting to be revealed.

1. What has been the hardest trial you have faced together? How did you handle it?
2. What are your personal tendencies during hard seasons or events (i.e., avoid it, talk about it, cry, and so on)?
3. Are you willing to ask your spouse for what you need during a trial, or would you be more concerned about meeting their needs? Decide now how you will handle pain together.
4. Being honest with your feelings plays a big part of enduring hard times. Do you have any concerns about sharing how you really feel with your spouse?
5. What is your biggest fear? Is this something you keep before the Lord in prayer? Take time right now to pray over this specific fear.

Dear Lord,

Thank you for being a God who is so big and so powerful. We cannot fully comprehend you or your plan for our lives. Thank you for being the hope that anchors our souls when times are hard. Help us to always trust in you and to never quit placing our hope in you.

In Jesus' Name, Amen

That's Not What I Said!

"What your eyes have seen do not hastily bring into court, for what will you do in the end, when your neighbor puts you to shame?" (PROVERBS 25:7-8).

A COURTROOM DRAMA might not be the setting to find encouragement for your marriage. The last thing you want to do is rush off to prove your case, only to find yourself lacking evidence and truth. You can assume you're right and still have no idea what's actually going on. Trust us, that's not what you're looking for in your marriage.

We've seen this drama unfold often. In fact, a lot of couples find themselves in our office because *what they think is happening* is not *what's really happening.* Even in small ways, assumptions can turn a simple situation upside-down.

Take Andrea and Austin, for example. Working from her laptop, Andrea would get frustrated with the slow Wi-Fi and technical difficulties. She'd vent "Why isn't this laptop working? This is so annoying!" Andrea feels frustrated with her laptop, but this isn't so clear to Austin, who takes her comment personally and responds defensively. He begins to drill her with accusatory questions and commands: "How many windows do you have open? When's the last time you turned it off? Quit clicking!" Already upset, Andrea doesn't take Austin's response well. Soon, they find themselves in an argument—neither of them really knowing what it's about.

What happened to Austin and Andrea happens to couples all the time. Austin heard something that wasn't said. He perceived an entirely different scenario than the one playing out. When Andrea expressed her frustration with the laptop, Austin heard her frustration *with him* for not fixing it—and he responded, as we all do, from his own perception.

When a conversation with your spouse suddenly goes a different direction than you predicted, or when you or your spouse seem to get upset "over nothing" or "out of nowhere," it's likely someone is responding to an assumption, not mutual understanding. And just like in the proverbial courtroom, you can come out of this looking a bit foolish.

If your spouse's response doesn't seem to fit the situation, stop and ask clarifying questions. Pause and ask your spouse what they heard you say or what they really mean, instead of assuming you know. Have an open conversation, and don't let unaddressed, false assumptions put your spouse on trial.

1. Have you ever fought over something seemingly insignificant? Is it possible a false assumption was behind the fight?
2. Some people are more prone to take things personally, make assumptions, or jump to conclusions. Talk about whether either of you fit in this category.
3. Think about how your tone might make it difficult for your spouse to understand your meaning. Are your words harsh, sharp, or snappy? Or are they soft and spoken with love? Based on Proverbs 15:1-2, why might your tone matter? How can you work on your tone to communicate your intention better?
4. What do you think the Proverbs verse above means when it says to not hastily bring these conflicts to court? Can you find other verses that add clarity to its meaning?
5. How can you make asking for clarification part of your communication process? What are some advantages of doing this?

Dear Lord,

Thank you for always being clear with your instructions and communication. Help us to learn from your ways. Show us how to not rush off in judgement or assumption, and teach us to be clear communicators and understanding listeners.

In Jesus' Name, Amen

The Big Cover-Up

"Above all, keep loving one another earnestly, since love covers a multitude of sins" (1 PETER 4:8).

IT'S NOT SURPRISING that Peter would remind us to love one another earnestly, or, as other translations say, "fervently." Peter was one of the most fervent lovers and followers of Christ, full of passionate intensity. He's the guy who cut off the ear of a man involved in Jesus' arrest, who boldly attempted to tell Jesus he knew the better plan, but also, when overwhelmed by fear, denied the one he loved and followed devotedly. Peter loved big and messed up big. Even still, Jesus loved him and showed him incredible grace.

When you love each other earnestly, your love invites grace to cover up any wrongdoings. In showing grace, God also shows us mercy by withholding from us what we really deserve for our sin, and instead, offering us his love. The grace we show one another should do likewise.

Hopefully, your spouse won't cut off anyone's ear, but it's pretty much a guarantee that they will make some mistakes. They will speak out of uncontrolled passion. They will disappoint you. They will not always be who you need them to be. Through grace, earnest love covers all of this. Just as you can't out-sin God's grace, your love should render your spouse unable to mess up too much, fall outside of grace, and lose your love.

Grace will always be unearned, because earned grace is not grace at all. However, even as God's grace covers our sin, it will not remove its natural consequences. In kindness, grace covers and teaches. The love shown in grace makes sin, poor decisions, and emotional outbursts much harder to commit without conviction. Grace does not desire you to repeat the offense or continue to walk in sin; it brings you to find freedom in its unearned richness, to learn from love, and to sin no more.

In your marriage, God asks you to do your part and earnestly love one another. Love each other so well that all the mistakes, failures, and disappointments will pale in comparison to the love and grace you hold for your spouse.

1. What do you think Peter's personality was like? Do you think you would have handled Peter with the same grace that Jesus did? Why or why not?
2. What are your thoughts on grace covering sin but not the consequences of sin? Does this make you feel better or worse? Why or why not?
3. How would you define grace? Does not having to earn grace ever make you feel guilty? How so?
4. What does grace look like when your spouse keeps making the same mistake over and over again? Would this verse still apply?
5. What do you think happens inside the heart of someone on the receiving end of grace? Do you think grace is powerful enough to result in change? How so?

Dear Lord,

Thank you for pouring out your love and grace on us, even in our sin. Thank you for allowing the blood of your Son to cover us in your grace for eternity. Help us to not try to out-sin the grace that we pour out on each other but to receive it with a grateful heart instead.

In Jesus' Name, Amen

Walking on Eggshells

*"There is no fear in love, but perfect love
casts out fear"* (1 JOHN 4:18).

HAVE YOU EVER FELT like you had to walk on eggshells around someone? Emma felt this way all the time around her husband, Wyatt, never knowing what might set him off. When he came home from work, she had to make sure their two little girls stayed quiet and everything in the house seemed perfect, or Wyatt would get upset.

In their marriage, Emma lacked the freedom to disagree, have a voice, and live without fear. If you can't disagree or share your thoughts with your spouse without upsetting them, love is bound to die off. Love does not exist where fear runs rampant. And Emma wasn't the only one missing out in their marriage.

Wyatt lacked the opportunity to learn and grow through seeing the consequences of his actions. Emma, in her fear of upsetting him, often hid the toll his behavior took on her. She tried to gain control of Wyatt's emotions by predicting and preventing his responses. Instead of allowing Wyatt to choose his own response, even if it was negative, she strained herself to stay a step ahead of him. Here's the deal—Wyatt getting upset is on him. And if he chooses unwisely, there should be consequences, which hold the potential to redirect his behavior. Instead, Emma continued walking on eggshells, and Wyatt continued to make bad decisions about his emotions, missing the opportunity to become more like Christ. Consequences are excellent teachers.

Emma and Wyatt's situation is a common one, and we've seen it play out across the spectrum, from everyday conflicts to abuse. The truth remains, "there is no fear in love." When you are free to disagree, you are free to love.

Real love accepts differing opinions, allows and respects being told "no," and embraces shared thoughts and feelings. Both of you need the freedom to feel upset and learn how to respond to your emotions. In your marriage, know where you end and where your spouse begins. It's okay to break a few eggs as you find out.

1. Do you ever hesitate sharing with your spouse because they might get upset? Why or why not?
2. How does it make you feel when your spouse gets upset over something you say or do? Why do you think it might not be good to take away their right to get upset on occasion?
3. What biblical examples can you think of where Jesus had boundaries or allowed others to be upset?
4. Have you ever agreed to do something you didn't want to do, because you felt bad about saying no, or worried that someone else might feel bad? Is this a common struggle for you?
5. In your own words, give each other permission to always be honest about your real feelings with one another.

Dear Lord,

Thank you for giving us permission to set boundaries, to say no, and to find freedom in our marriage, as your perfect love casts out fear. Help us to walk in self-control and to learn from the consequences you allow in our lives. Help us to always have the courage to be completely honest with one another and without fear of one another's responses.

In Jesus' Name, Amen

A Time to Grow

"And so, from the day we heard, we have not ceased to pray for you, asking that you may be filled with the knowledge of his will in all spiritual wisdom and understanding, so as to walk in a manner worthy of the Lord, fully pleasing to him: bearing fruit in every good work and increasing in the knowledge of God" (COLOSSIANS 1:9-10).

HAVING A GOD-CENTERED MARRIAGE requires knowing God. And meeting God, acquainting yourselves with the idea of doing marriage his way, is just the beginning. If you really want him as the focus of your relationship, you must *really* get to know his heart, his character, and his desire for your lives. As you grow to know each other more, keep getting to know him more, too.

What would your marriage be like if you never tried to figure out the likes and dislikes of your spouse? What if you never cared enough to ask for your spouse's opinion? Clearly, a relationship like that would be doomed and stagnant. The same can be applied to your relationship with God. Having an authentic, God-centered marriage means focusing on your relationship with God, growing spiritually as individuals and as a couple. You are each responsible for your relationship with God, and the health of that relationship makes a huge difference in the health of your marriage.

Sophia constantly struggled with insecurity. For years, this affected her marriage with Brett. She relied on him to help encourage her almost every day, sharing her fears and discouragement over and over again—but her problems were way too big for Brett to solve on his own. What Sophia really needed was God's wisdom and help. In addition to feeling stuck in this conversation, Brett and Sophia couldn't even hang out with friends anymore, because Sophia's insecurities increased around them.

Finally, with Brett's encouragement, Sophia decided to take her struggles to the Lord. She started attending a weekly Bible study with other women, digging into Scripture, and writing in a prayer journal, where she asked God for his help. Slowly but surely, her insecurities started to dwindle, as she gained a bigger community and a wider perspective of God and her identity in him. Brett began to see the fruit of her newfound security, and they progressed in their conversations and enjoyed time with friends again. By focusing on her relationship with God first, Sophia found more security, and Brett found himself less defeated by his limitations to help his wife.

As long as you walk this earth, God desires for you to walk in a manner worthy of him. Grow and bear fruit, investing in your relationship and knowledge of God. Move closer toward God daily, and you will find yourselves moving closer to one another.

1. If neither of you desired to grow spiritually, what would happen to your marriage?
2. How are you currently investing in your individual relationships with God?
3. Is there a specific area where you lack spiritual growth? What would an investment in that growth look like?
4. Is there a Bible study or a sermon series you could go through together? Discuss what topics you might be interested in learning.
5. Are there groups you could join to grow spiritually with friends (i.e., marriage ministry, Bible study, home group, etc.)? Discuss your ideas and how to make them happen.

Dear Lord,

Thank you for desiring us to walk with you and to learn your ways. You alone know what is best for us and the best way to live out our marriage. Help us to never stop pursuing knowledge of you, and help us apply all that we learn together.

In Jesus' Name, Amen

It's a Date!

"Enjoy life with the wife whom you love, all the days of your vain life that he has given you under the sun, because that is your portion in life and in your toil at which you toil under the sun" (ECCLESIASTES 9:9).

YOUR MARRIAGE will certainly take hard work, but you should also find marriage enjoyable. After all, your marriage's primary purpose is to represent the relationship between Christ and his bride, the Church. And how many people would be drawn to a relationship with Jesus, if its model just seemed heavy, complicated, and hard all the time? Most people likely have relationships like that in their life already.

Marriage takes work, but it's also fulfilling and fun. It should have rewards, celebrations, and laughter. Sometimes, when marriage does get hard, it's because you've forgotten it can be fun. Perhaps you've neglected to enjoy your spouse—spending time together, investing in one another, relaxing side-by-side, and simply having a good time.

How can you enjoy your marriage, even when life gets tough or busy? Schedule regular date nights. Sean and Olivia have managed to prioritize dating one another, even as a married couple. Not every date night is expensive or even outside of their house. Each week, they sit down with their schedules and decide when to spend time together. From coffee dates in the kitchen before work to Friday night concerts, they intentionally make time to enjoy one another's company.

Date nights, or date days, don't have to be complicated. On occasion, elaborate dates are fun and romantic, but you can enjoy one another with simplicity. You don't need to reinvent the dating wheel; let Pinterest plan your date nights. Look up and share creative ideas to invest in your marital joy, many of which won't cost you a dime.

Enjoying life together matters. Many pursuits in life may be done in vain, but spending time on your marriage is not one of them. Your relationship will always be worth it. Your most fun dating life still lies ahead of you.

1. Talk about the last date you went on. Did you enjoy it? Is it something you would do again?
2. What are your date expectations? Do you want something elaborate, or will a coffee date suffice? Does it matter to you who plans the date? Be honest with your expectations and desires.
3. Do you have similar date expectations? Do you have a fear of disappointing your spouse, if you plan a date they don't like?
4. Time goes by fast, and God wants you to enjoy each other (Ecclesiastes 9:9). How can you mix up your dating routine to make it enjoyable? Is there a way to be more creative with date nights? Is there a regular day of the week you can schedule them, so that you have less opportunity to skip them?
5. Take five minutes to look up date-night ideas online. Choose one you can do this week that you will both enjoy.

Dear Lord,

Thank you for giving us each other to enjoy life with and have fun together. Help us to not waste our time living a life in vain but to live in a way that draws others to you. Let our fun and our connection allow others to see how life with you is better.

In Jesus' Name, Amen

Continuing the Journey

"Unless the LORD builds the house, those who build it labor in vain" (PSALM 127:1).

THIS IS WHERE we come full circle. It's time for your God-centered marriage to continue along a journey of great purpose. And the only way to live out a purposeful marriage, pleasing to the Lord, is to surrender yourselves to him. As a couple, focus entirely on God throughout your lives together. Allow him to build your marriage in the way he desires.

We can't imagine ending this book any other way than sharing one final insight in regard to our friends, Ryan and Beth. Ryan loves his wife like Christ loves the Church, and Beth respects her husband tremendously. The two of them display the gospel in their marriage so well, and it's evident to anyone who gets to know them.

One of the reasons Ryan and Beth have this kind of marriage is because they are intentional. Early on, they decided together that their marriage and family life would revolve around loving God and loving others. It's the core of who they are and how they coexist together. It's a filter for their lives. When deciding what to do about anything that takes their time or resources, they ask themselves, "Does it help us love God and love others?" If not, it will hold little to no place in their life. But, if it does, consider it done.

Loving God and loving others is simply who they are. They both eat, sleep, and breathe this love, and their mission identifies and solidifies them as a couple. It places God in the very center of their marriage and holds him there resolutely, for today and all the days to come.

As this devotional comes to an end, we want you to achieve what this exemplary couple has, and share in their joy and connection with the Holy Spirit. Have a core value you hold as a couple that identifies who you are and answers God's calling in your life. Let this love add meaning to your vows every day. At the mention of your names, may the Name of Jesus always be heard louder.

1. Think about couples you know. Do you know a *Ryan and Beth*—a couple whose life clearly reveals their purpose?
2. How could your mission statement serve as a filter for your marriage? How could it affect your schedule, your commitments, and your decisions?
3. What have you learned from all of these weeks of devotions together? Has placing God in the center of your marriage made a difference? Discuss this together.
4. What do you want to do next as a couple to continue keeping God in the center of your lives? What are you most excited about in the year ahead?
5. What do you want your core value to be as a couple? Spend time thinking about this, then make your marriage mission statement.

Dear Lord,

Thank you for being the center of our marriage and the builder of our home. More than anything else, we desire for your Name, Jesus, to be known, for it is the only name that can truly save. Together, we ask that you bless our union. Make use of our marriage to bring you glory and expand your kingdom beyond what greatness we could ever imagine.

In Jesus' Name, Amen

References

The Holy Bible, English Standard Version. ESV® Text Edition: 2016. Copyright © 2001 by Crossway Bibles, a publishing ministry of Good News Publishers.

"The Lone Barone," *Everybody Loves Raymond* (1998). (Referenced in Week 9, "For Better or For Worse.")

Ramsey, Dave, *The Total Money Makeover: A Proven Plan for Financial Fitness* (Nashville: Thomas Nelson, 2003). (Referenced in Week 19, "The Value of Things.")

Ramsey, Dave, Financial Peace University (www.daveramsey.com). (Referenced in Week 19, "The Value of Things.")

Gottman, John, The Gottman Institute (www.gottman.com). (Referenced in Week 23, "Can't We Just Get Along?")

Smalley, Gary, *The 5 Love Languages* (www.5lovelanguages.com). (Referenced in Week 27, "The Languages of Love.")

Welch, Kristen, *Made to Move Mountains: How God Uses Our Dreams and Disasters to Accomplish the Impossible* (Grand Rapids: Baker Books, 2020), 40-41. (Referenced in Week 30, "Made for More.")

Feldhahn, Shaunti, *The Surprising Secrets of Highly Happy Marriages: The Little Things That Make a Big Difference* (Colorado Springs: Multnomah Books, 2013), 116-117. (Referenced in Week 35, "What Are You Waiting For?")

Resources

Boundaries in Marriage, Dr. Henry Cloud and Dr. John Townsend

Boundaries in Marriage Workbook, Dr. Henry Cloud and Dr. John Townsend

The 5 Love Languages, Gary Chapman (www.5lovelanguages.com)

How We Love, Milan and Kay Yerkovich (www.howwelove.com)

Financial Peace University, Dave Ramsey (www.daveramsey.com)

The Total Money Makeover, Dave Ramsey (www.daveramsey.com)

SYMBIS: Saving Your Marriage Before It Starts, Drs. Les and Leslie Parrott (www.symbis.com)

Acknowledgments

WITHOUT THE GRACE AND REDEMPTION OF GOD, this book would have never been possible. He will always have our appreciation first—for always showing up in our lives, redeeming all the areas we blew it, and using everything the Enemy meant for harm for good. Thank you to each and every couple who invited us into their story, by meeting with us in person or by picking up this book. Your desire to strive for better in your marriage keeps pushing us forward.

Thank you to our daughters. First, Alicia, for being the one who kept us on our deadline and pushed us to write without overthinking, cheering us on the whole way. Taylor, for always interrupting with her phone calls that just so happened to provide us with the breaks we needed, and for always thinking and telling us that we are the best in the universe. Finally, to Mackenzie, for being the English and psychology scholar who taught us all the proper placements of every single comma in this book. Thank you all for always believing in our ability to write and to change lives with the truth of the gospel. We truly hope that the writing of this book encourages each of you to accomplish all that God has called you to.

Lastly, thank you to our team of editors, Kim Suarez, Meg Ilasco, and Susan Randol, who were always there to answer our questions, reassure us, and encourage us along the way.

About the Authors

CHRIS AND JAMIE BAILEY are professional Christian counselors and marriage coaches. They have been married for over twenty-five years and, together, raised three daughters. Everything they teach other couples was learned firsthand through their own backgrounds and marriage. They run a private practice in South Carolina as well as Expedition Marriage, an online marriage ministry. They are certified facilitators of Prepare and Enrich and SYMBIS, two of the top premarital and marital assessment programs for couples. Through speaking, writing, and hosting marriage retreats, they desire to share the truth of God's Word, along with practical tools for Christian couples everywhere to live an abundant life through Jesus. Learn more and access their online resources at expeditionmarriage.org.

NOTES

NOTES

NOTES

NOTES

NOTES

NOTES

NOTES

Hi there,

We hope you enjoyed reading *Newlywed Couple's Devotional.*
If you have any questions or concerns about your book, please
contact **customerservice@penguinrandomhouse.com**
so we can take care of them. We're here and happy to help.

Also, please consider writing a review on your favorite retailer's
website to let others know what you thought of the book and to
help them with their buying decision.

Sincerely,
Zeitgeist Publishing